Praise for *The Engineering Leader*

Cate is an expert on managing and leading whose work I recommend again and again. Her new book is a wealth of great advice, helpful insights, and useful tools, and will improve leaders, their teams, their cultures, and their companies.

—*Ellen Pao, tech investor, CEO of Project Include, and author of* Reset

This book is a valuable guide to navigating career growth and making an impact in any role, whether you lead teams, write code, or are figuring out which path you want to take.

—*Camille Fournier, author of* The Manager's Path

This book will give you a playbook for specific situations you might find yourself in—but more importantly, it will give you the tools, muscle and confidence to handle whatever else your leadership role throws at you.

—*Tanya Reilly, author of* The Staff Engineer's Path

I've heard advice from many brilliant people over my career, but few of them have had the impact that Cate has had. Her advice helped to form my team, my company, and my own leadership style. This book has everything she taught me, and I cannot recommend it highly enough.

—*Dan Shapiro, CEO of Glowforge, Inc.*

The Engineering Leader is packed full of practical advice and frameworks that will not only help you better manage your team, but also better manage your own energy and your career moves—topics that books on management often overlook.

—*Jill Wetzler, VP engineering and consultant*

An essential guide for tech managers, combining personal experiences with practical strategies. This book is essential for anyone entering the challenging but rewarding field of technology leadership.

—*Taylor Dolezal, head of ecosystem, CNCF*

The Engineering Leader

Strategies for Scaling Teams and Yourself

Cate Huston

Beijing · Boston · Farnham · Sebastopol · Tokyo O'REILLY®

The Engineering Leader

by Cate Huston

Copyright © 2024 Cate Huston. All rights reserved.

Published by O'Reilly Media, Inc., 1005 Gravenstein Highway North, Sebastopol, CA 95472.

O'Reilly books may be purchased for educational, business, or sales promotional use. Online editions are also available for most titles (*http://oreilly.com*). For more information, contact our corporate/institutional sales department: 800-998-9938 or *corporate@oreilly.com*.

Acquisitions Editor: David Michelson	**Indexer:** Judith McConville
Development Editor: Michele Cronin	**Interior Designer:** Monica Kamsvaag
Production Editor: Elizabeth Faerm	**Cover Designer:** Susan Thompson
Copyeditor: Shannon Turlington	**Illustrator:** Kate Dullea
Proofreader: Piper Editorial Consulting, LLC	

April 2024: First Edition

Revision History for the First Edition

2024-04-16: First Release

See *http://oreilly.com/catalog/errata.csp?isbn=9781098154066* for release details.

The O'Reilly logo is a registered trademark of O'Reilly Media, Inc. *The Engineering Leader*, the cover image, and related trade dress are trademarks of O'Reilly Media, Inc.

The views expressed in this work are those of the author and do not represent the publisher's views. While the publisher and the author have used good faith efforts to ensure that the information and instructions contained in this work are accurate, the publisher and the author disclaim all responsibility for errors or omissions, including without limitation responsibility for damages resulting from the use of or reliance on this work. Use of the information and instructions contained in this work is at your own risk. If any code samples or other technology this work contains or describes is subject to open source licenses or the intellectual property rights of others, it is your responsibility to ensure that your use thereof complies with such licenses and/or rights.

978-1-098-15406-6

[LSI]

Contents

Foreword

Leadership roles come with a lot of *shoulds*. Read any leadership book (and my own is no exception), and you'll find a ton of emphasis on how important it is to do the job right and the many, many ways you can do it wrong. When you first become a people manager, a Staff+ engineer, or any other kind of engineering leader, you might be surprised by how much is expected of you. If you're a manager, you have responsibilities toward the people who report to you, of course, but any leadership role also means obligations toward users, colleagues, business goals, the company's bottom line, and, depending on what you're working on, maybe the society you live in. Maybe the whole planet. The expectations are high, and that's fair. As an engineering leader, you can solve huge problems and shape the industry for the better, or you can create *absolute chaos.*

It sometimes feels like *everyone* has a story about a leadership failure they're still upset about. You'll hear about managers who burned out their teams with unmeetable goals or changed direction so often that the Big Exciting Project never shipped. Maybe it's a Staff+ engineer who saw themselves as the gatekeeper of their codebase or architecture and who defended that gate with all their might. (Their preferred weapons: withering scorn, unmeetable standards, refusal to help. Other engineers eventually stop trying.) It could be a leader who meant well but avoided conflict, or who wouldn't make decisions, or was completely absent, or overshadowed the team so nobody else could grow. And of course you'll hear about the micromanagers, the blamers, the credit-stealers, the microaggressors, the...ok, like I said, there are a lot of ways to get it wrong.

But there are also a lot of ways to get it right. You'll hear good stories too: lead engineers who shipped an incredible product, bringing their colleagues up a level while they did it; managers who led their teams out of reactiveness and toil and into sustainable, predictable execution; role models who created environments where everyone could do great work. You'll hear people credit "a great manager I had at the time..." as the starting point or turning point of their rocketship career. And strong leaders can define the culture, values, and ethical standards for their whole organization. So, whether on the management or technical/IC track, anyone in a leadership role can be an incredible force for good. The expectations are high because it matters. The stakes are high. The problem is, the difficulty level is high, too.

As an engineering leader, you'll wear a lot of hats: you're the coach, the strategist, the project lead, the on-message agent of the company, the supporter, the adult in the room, the planner, the incident commander, the decider, the umbrella, the resilient and emotionally mature person who (somehow?) knows the right thing to say in every difficult situation. Not all of those roles will come naturally. Being the lead can be overwhelming, but you're not supposed to get overwhelmed. In fact, you're expected to show up with enough excess capacity to bolster everyone else, too. When the project is daunting, it's your job to project confidence and help everyone else believe in the mission. And when other people depend on you, it's much harder to disconnect: sure you can set a good example and take a mental health day...so long as you reschedule eight meetings first. As Cate tells us in Chapter 6, "It's so easy to get caught up in what other people expect of us and put ourselves last." It's all about the team, right?

That's what I love about this book. In *The Engineering Leader*, Cate doesn't insist that you put yourself last. In fact, just the opposite: rather than opening with expectations and obligations and all of those *shoulds*, the book starts by asking what *you* need. This inner focus will be a new idea for a lot of readers who for years have been hitting snooze on their own career introspection "just until things get a little less busy." But Cate knows that you can't lead other people unless you're in good shape yourself, so the book opens by inviting you to take a hard look at what you're trying to achieve for yourself and your career. There aren't any "right" answers here but rather a focus on understanding what your options are and figuring out what opportunities you're hoping to unlock. With that direction in mind, Cate invites you to think about how you work and whether you'll need to build new energy management strategies or expand the techniques in your leadership range to get to where you want to be.

In fact, the whole first half of the book is about *you*, the leader, setting yourself up for success, building skills and structures for self-management so that you can focus on what you need to deliver: a high-functioning team that gets the work done. And that's the second half of the book. "Accept that managers are overhead," Cate says—and then explains how to make that overhead worthwhile. Let's be clear, this is a book that believes in feelings, and it checks in with those feelings and respects what they have to say. But for all the attention to feelings, it's never idealistic; it acknowledges the reality that engineering teams exist for a purpose, and leaders are there to deliver on that purpose. In Cate's worldview, there's no contradiction there: everyone gets to feel how they feel and then you all go get sh*t done.

In this book, you'll learn how to build healthy teams that are delivering business goals. You'll scale those teams, and scale yourself too, by building in peer support and ecosystems that help everyone succeed. You'll understand the *why* of the work, so you can prioritize which things you *have to* get right and which can be just good enough. And you'll learn how to introduce processes that colleagues are enthusiastic about because they help them move faster, and not... the other kind. (The penguin example alone is worth the price of this book. I'm not even kidding.)

Some of the skills Cate teaches are needed by any leader: managers, Staff+ engineers, and even senior ICs will get a lot from the sections on how to conduct a technical interview, how to give feedback, how to delegate and make a problem tractable for a less experienced colleague, how to contribute to an environment where everyone can thrive, how to debug a blocked team and get things moving again. Other topics, like hiring, engineering metrics, or performance management, are primarily for managers, but the rest of us can always benefit from seeing what's going on behind the scenes there too.

"Congratulations! You are now the manager of yourself," Cate says, before reminding us, "Wait...this was always true." Like a good coach, Cate unlocks potential and builds capabilities rather than just handing out advice. This book will give you a playbook for specific situations you might find yourself in, but more importantly, it will give you the tools, muscle, and confidence to handle whatever else your leadership role throws at you. And the focus on managing yourself in a sustainable way means you'll be able to fulfill the (again, frustratingly valid!) high expectations of the role.

I took so many notes while reading this book: some philosophy, some concrete techniques, and, honestly, just some zippy one-liners. (I love when Cate becomes, as she describes it, "icily British.") I hope you enjoy it as much as I did.

—*Tanya Reilly*
Senior Principal Engineer and
author of The Staff Engineer's Path

Preface

Early on in my career, my goal was to become a staff engineer. I loved programming. I loved the high I got each time I figured something out, the thrill of progress each time I merged a pull request (PR). I loved shipping things. I loved it when people used things I'd built.

But four years out of grad school, navigating multiple reorgs and enduring dysfunctional teams that didn't ship, I left that job. I realized that it didn't matter if my job sounded good if I didn't *feel* good about it, and that growth and learning were not synonymous with getting promoted. I took a break and tried to reconnect with that love of programming again. Worked on my personal projects and finally wrote an algorithm and optimized it for space and time outside the context of an interview.[1]

It turned out that the most easily monetizable skills were actually the soft skills—the things I described as "thankless emotional labor" when my job was programming. My career inverted. Instead of spending the workday programming, I got paid to do technical interviews and help startups build up their development teams. I wrote code for fun, turning my little side projects into a technical book chapter and an app on the app store.

After I spent a year rediscovering my love of programming and learning the value of nonprogramming skills on the open market, a friend recruited me to manage a team in a remote-first company. I shipped out to Colombia and started managing a small team. However, the subsidiary and visa I had expected never materialized, the startup imploded, and I wondered if I would go back to being a developer again.

---

1 This is a joke about how we interview programmers—and how often interviews force someone to demonstrate they can optimize algorithms, something that rarely happens on the job.

I didn't. Somehow, I acquired another job and a team four times the size of the one I managed at the startup, and I navigated the learning curve to manage managers. This time, the team was fully distributed—back then, this was still far from the norm—and role models and advice were all presented in an on-site office context.

Two things stood out to me in that first year of management. The first was that being a manager could be crushingly lonely. Especially at a startup, with no real peers.

The second was the difficulty of building a model of what to go toward—something that became all the more stark when I started managing managers. I felt like I'd barely got a handle on how to manage a team.

During that time, I thought a lot about the worst manager I had ever had. I remembered the time when during a 1:1 he explained to me that *his* manager had been terrible, and so he must be doing better because he wasn't *that guy*. And I realized many managers have no better model than "don't be that guy." I was lucky to experience up close how little a mindset like that benefited a manager, and I was determined to avoid that mistake myself. I realized the importance of building something to go toward, and how little I had to go on as to what that would look like. That I had never been part of a functional team, really. Never had a great manager, and never even had a good manager for very long, given the annual reorgs and lack of continuity in my management chain. That I had found the most growth, learning, and support in the community rather than in the workplace.

Defining this model of what to go toward as a leader gave me something to live up to, even if I sometimes failed. Keeping at it, after the first trial by fire, gave me the opportunity to give other people what I had never had. I was able to reach down and pull others up behind me and give them an easier time of it than I had myself. I got the opportunity to experience leading a functional, high-performing team, beyond what I got to be part of as an individual contributor. A great team, a supportive and productive environment where people can learn and grow and *ship* is both more intransigent and more permanent than the software we create. Teams, like products, get discontinued; for example, the team at the defunct startup I worked at will never exist in that configuration again. But the impact on *people* remains, and this compounds even as their careers continue. This is the impact that a good team can have.

This book is the result of my attempts to build that model. If you've been lucky enough to have a good manager, perhaps it will help you develop more depth and understanding about what goes into that. But if you've never had a good manager, or don't currently feel like you *have* a manager, know that I wrote this book for you. On the days that you feel lost, or unsupported, or overwhelmed, I want you to know that these feelings are normal and that there *are* things you can do.

Every time I talk to an anxious new manager, I give them the same advice. I tell them they will get a long way if they consistently show up and give a damn about the people they manage as human beings. I believe that is 80% of the work. We all know, though, that the last 20% is where the problems lie. I hope this book will help you navigate the last 20%.

Who the Book Is For

This book is for two main types of people. It's for leaders—folks who manage engineering teams. It's also for individual contributors (ICs)—the folks who do the day-to-day work of writing code, shipping products, and so on. Whether you're in a leadership role or an IC role, whichever way you slice it, you're still the manager of yourself. The guidance in this book is written to help you navigate your career, your organization, your team—and to make the best of what you have around you, whatever that is, in the different forms it comes in.

For managers, this book is my attempt to help you find that model of what "good" leadership looks like, finding something to go toward—something beyond the tactical list of things you're responsible for or the tasks you're expected to accomplish. I encourage you to develop that model in terms of not only what it means to lead a team but also—perhaps even more important—what career growth even means and how to find it independent of job titles and leveling frameworks.

For ICs, understanding more about how things work can be invaluable for identifying what questions to ask, where to exert pressure, and what kind of constructive feedback to give. The first half of the book was written to be applicable to you as an individual and as a leader, and the second half will help you gain context to better understand what goes on around you, both on your team and in the larger context of the organization and your career.

In my coaching practice, I often find that ICs lack the context that places their frustrations in perspective; the more I rise through the org chart and understand a greater context, the more I wish I could revisit my past-Cate self and help her understand this. My hope is to provide ICs with more of this perspective that will help you gain additional insights to navigate challenges and grow your career more effectively.

Even if you are lucky enough to have a manager who does care about your career growth and tries to help, your career will be bigger than your current role. Your goals may eventually come into conflict with what your current role can offer you, and you will need to decide what to do about that. Developing your own point of view and a broader base of support can make you more resilient to the inevitable reorgs, layoffs, and disappointing review cycles.

Organization of the Book

This book is organized into two parts, with two sections in each:

- Part I focuses on *you*—what do you need to be effective and fulfilled?
    - Section 1, "DRIing Your Career" focuses on your growth and goals, including your relationship with feedback.
    - Section 2, "Self-Management" aims to answer the question "What is my job, and how do I get better at it?"
- Part II focuses on *the team*—how do teams scale, and how do they deliver?
    - Section 3, "Scaling Teams" is about the people management practices—like hiring, developing people, and firing—that are required, particularly when navigating periods of growth.
    - Section 4, "Self-Improving Teams" to Be Self-Improving is about the layers of team functioning—mission, strategy, tactics, and execution—and how to drive improvement to them.

Section 1, "DRIing Your Career" is about career development and what it means to be the directly responsible individual (DRI) of your career. It is for ICs who need a framework for thinking about their career in a more expansive way than their current role, and for leaders who not only need to do that for themselves but also help others with that same problem. It is not a plan or a prescription; it is a model and way of thinking that you can use both in your own career and as you support the people whose careers you have impact on. We'll also talk about feedback: how to get it, how to make the most out of it, and when to put it in the bin. This section is to help you think critically about what you want from your career and how to get that. It comes first because it is foundational. Because no one will do this work for you, and you will struggle to help anyone else with this until you have done it for yourself.

Section 2, "Self-Management" covers the concept of self-management, the work you do to self-regulate and define your role such that you can be an effective leader for your team. It is for those navigating the sudden shock of the leadership role—whether as a first-time manager or in a Staff+ role—and for those who need to revisit some of the ideas as they take on more responsibility and challenges. The struggle many well-intentioned, nice managers have is that they subsume themselves to the needs of the team, prioritize happiness over effectiveness, forget the big picture in the minutiae of the day-to-day. This leaves them burned out and ineffective or some combination of the two. If you don't have a good manager helping you, self-management helps you avoid this trap. If you do, self-management helps you help yourself and creates space for your manager to support you with more impactful things.

Section 3, "Scaling Teams" is about how teams scale. This section is for managers, although the context may also be useful to senior ICs wanting to understand why things are the way they are and what can be better. Scaling is not just adding more people; it is evolving processes and structure such that you can add effective people. It covers how to hire effectively while improving diversity, effective onboarding, supporting and developing people, and navigating the challenges of performance management.

Section 4, "Self-Improving Teams" is about the impact of the team—the output of what you build together—and is for both (senior) managers and technical leaders (Staff+). Teams exist in service of something, and teams that do not deliver value do not survive—but it is possible to drive outcomes while caring about people as human beings. At the core of this is the idea that teams communicate in layers: mission (or vision)—the what, strategy—the how, tactics—the observability and interoperability of teams, and execution—the getting sh*t done. Understanding these layers helps you identify how to create clarity, improve effectiveness, and set a team up to be self-improving.

At the end of each section is an action plan for you to apply to your current situation, and throughout you will find reflection prompts: questions for you to take some time to consider or discuss with a peer, a friend, or a coach.

A core idea in developing strategy is that good strategy is *contextual*. As such, your strategy for your current challenges—whether about your career, your current role, or your team—depends on your context. My hope is that the action plans and reflection prompts throughout this book will help you define, and refine, *your* strategy, and that this strategy supports both your own well-being and the effectiveness of the team you lead.

O'Reilly Online Learning

O'REILLY® For more than 40 years, *O'Reilly Media* has provided technology and business training, knowledge, and insight to help companies succeed.

Our unique network of experts and innovators share their knowledge and expertise through books, articles, and our online learning platform. O'Reilly's online learning platform gives you on-demand access to live training courses, in-depth learning paths, interactive coding environments, and a vast collection of text and video from O'Reilly and 200+ other publishers. For more information, visit *https://oreilly.com*.

How to Contact Us

Please address comments and questions concerning this book to the publisher:

O'Reilly Media, Inc.

1005 Gravenstein Highway North

Sebastopol, CA 95472

800-889-8969 (in the United States or Canada)

707-827-7019 (international or local)

707-829-0104 (fax)

support@oreilly.com

https://www.oreilly.com/about/contact.html

We have a web page for this book, where we list errata, examples, and any additional information. You can access this page at *https://oreil.ly/engineering-ldr*.

For news and information about our books and courses, visit *https://oreilly.com*.

Find us on LinkedIn: *https://linkedin.com/company/oreilly-media*.

Watch us on YouTube: *https://youtube.com/oreillymedia*.

Acknowledgements

My dear friend Dr. Linda Carson passed shortly before I started writing this book. Linda was a polymath: she held degrees in mathematics, kinesiology, fine arts, and psychology and was also an MFA. She was a leader in interdisciplinary learning, developing the curriculum for the University of Waterloo's Knowledge Integration program. I was lucky enough to know her when I was living in Kitchener-Waterloo in Canada, shortly after escaping (aka dropping out of) grad school to work at Google. Linda taught me many things. She inspired me to think about art in a different way, influenced my thinking about women in the workplace (and beyond), broadened my thinking about narratives and tying seemingly disparate things together, and introduced me to all the best food in that area. She pushed me to think critically, challenged me to balance my ambition with some modicum of self-care, and always had a wise or supportive word when I needed it. I still think about things she taught me, and I miss her deeply. I like to think she would be proud of what I've written here, but I think above all she'd be proud that I *did the thing*. Her passing, too young, is a constant reminder to me that life is short and you get one, so chase the meaning.

I'm so grateful to O'Reilly Media for publishing my book. Melissa Duffield rescued me from months of procrastinating on writing a book proposal (without her I might still be "working" on it). David Michelson, the acquisitions editor who guided the book through the process. Michele Cronin was a tremendous editor: insightful, supportive, and *incredibly* helpful. Liz Faerm handled productionization of the book into the finished product. Shannon Turlington diligently copyedited everything. Susan Thompson designed the beautiful cover.

Camille Fournier and Tanya Reilly, thank you for taking this path before me, leaving an easier one for me to follow. Thank you for all the encouragement every time I felt like writing a book was *particularly* hard, and for every time you've cheered me on. Camille, I'm additionally grateful for more than a decade of friendship and adventures on multiple continents. Here's to many more! Tanya, thank you for writing the most beautiful foreword and for all your behind-the-scenes snippets of advice.

My friends Nandana Dutt and Eli Budelli have been such a support in this adventure. Nandana, I am grateful for all the ways that you have broadened my thinking by sharing yours, and, of course, for all the pep talks. I'm so happy to have your case study enrich the book. Thank you. Eli, thank you for always believing in me, even (especially!) when I don't believe in myself, and for telling me everything as it is. As well as your story (thank you!), your insight is scattered throughout the book. You will always be my best work BFF, although I will never do manual labor on your vineyard. I love you both.

My friend Jill Wetzler contributed a case study and a story and was also the most diligent and thorough reviewer. Jill, this book is much better for your contributions, and I appreciate you so much. Thank you.

It was really important to me to include more perspectives in this book other than my own, and I'm deeply grateful to Lore Mattei, Jean Hsu, and Yuelin Li for their case studies, and to Alexander Coffman, Ellen Chisa, Jazbo Beason, Julie Pagano, and Margarita Gutierrez for the generous personal stories they shared.

My friend Hilary Mason was probably the first person to believe I could write a book. I was several years behind her, but who knows how long it would have taken without her encouragement? Clive Thompson took the time to talk to me about the practicalities of publishing, which was tremendously helpful. Heather Landy edited so many of my pieces for Quartz and showed me how to be a better writer. Ellen Pao's interview structure for the Project Include Remote Work report (*https://oreil.ly/ei1OO*) and comments about managers in a remote context stuck in my head and sparked something; that conversation was the catalyst for Section 3, "Scaling Teams". Dan Shapiro supported my ideas about hiring better to the point that he let me test them on his startup; this work laid the foundation for Chapter 8. Matt LeMay gave me the perfect pep talk at a time when I was in dire need of it, along with some very helpful advice. Chiu-Ki Chan helped me get into public speaking and ran the *Technically Speaking* newsletter with me for more than three years; many of the ideas in this book started out as talks, and the talks would not have happened without Chiu-Ki.

At many points in this book, I extoll the value of coaching, a testament to the coaches I have worked with who have been so incredibly helpful. My speaker coach, Denise Graveline, inspired me with her Eloquent Woman blog and taught me about metaphor and narratives. Denise passed in 2018, but I continue to carry and apply everything she taught me. Alison Jones of Practical Inspiration Publishing was an incredible help turning my idea into a structure and a book proposal. Dani Rukin, my first and longtime coach, helped me navigate some of the hardest parts of my career and got me into PQ. Jitka Peeters, my current coach, has been an invaluable support trying to balance writing a book alongside working a demanding job. Duncan Skelton, whose coaching I experience vicariously via some of my directs, continually shows me the impact a great coach can have on the people they work with.

Gratitude to my coworkers Aitor Viana Sanchez, Anirvan Chakraborty, Clodagh McCarthy Luddy, Bartek Waresiak, Elle Sullivan, Marcos Holgado, Mark Ingram, Nada Aldahleh, and Zbig Motak. Particular gratitude to my boss, Caine Tigne, who both supports and challenges me, and to Diana Kalkoul, my friend and recruiter at three (!) companies who taught me so much about hiring well and scaling.

I appreciate the time and insight from technical reviewers Anirvan Chakraborty, Jill Wetzler, Kamilah Taylor, Kaya Thomas, Kevin Stewart, Taylor Dolezal, Ian Nowland, Jesse Anderson, and Marcus Blankenship.

A surely incomplete list of the many people who have taught me things and influenced my thinking: Akshay Shah, Alan Eustace, Anne-Marie Imafidon, Ashe Dryden, Beau Lebens, Blake Janelle, Camille Emma Acey, Carla Geisser, Catt Small, Dave Stewart, Diana Kimball Berlin, Duretti Hirpa, Erica Joy Astrella, Jamie Kelly, Jo Miller, Jocelyn Goldfein, Juan Pablo Buritica, Judith Michelle Williams, Julia Ferraioli, Karen Catrin, Kelly Rusk, Kendra Carattini, Leigh Honeywell, Lena Reinhard, Leslie Ikemoto, Maggie Zhou, Marcellus Mindel, Marco Rogers, Martin Pilkington, Mary Selby, Matt Mullenweg, Matthew Miklic, Mekka Okereke, Melissa Dominguez, Meri Williams, Michelle Broderick, Mike Perrow, Minnku Song, Natasha Murashev, Nicole Sanchez, Nigel Babu, Nikolay Bachiyski, Olivia Shen Green, Pearl Latteier, Priyangee Guha, Ramya Sharma, Sophie Roberts, and Tracy Chou.

Thank you to our housekeepers, Paul and Monika, without whom we (or at least I) would be feral, and to my parents, Paul and Helen.

Finally, to my partner, Bas, who has been both a practical and emotional support as I wrote this book and is endlessly understanding about the amount of time I shut myself away to focus and write. I love you so much; thank you for making me feel both held and free.

PART I

You

Often when we talk about teams and individuals, these things seem divorced from each other. When we talk about teams, we talk about belonging to *one* team; the individual is subsumed, sacrificing for the team glorified. And yet when we talk about individuals, we emphasize the single person and ignore all the other people whose contributions were vital. Neither of these is realistic. For individuals, teams are transient, coming and going, and they may belong to more than one team at any given time—at work, the team they lead and the team of their peers, and outside work, the team of open source software contributors or the sports team. On all teams we're part of, we (hopefully) agree more than we disagree—but that doesn't mean we agree on everything. And when individuals move on, for whatever reason, (mostly) the team continues.

The reality is that teams are made up of individuals, and they all have their own wants, needs, and desires. That includes the leaders of the team—aka, you. The first half of the book is all about giving yourself space for your needs: what you need as the DRI (directly responsible individual) of your career and what you need to manage yourself so that you can manage your team effectively (or decide that management is not for you at this time or in this job).

This might be an uncomfortable concept. Sometimes, nice people have the idea that any consideration of their own needs is selfish and wrong. But there is a wide expanse between awful managers who happen to be selfish and ineffective managers who resent that implication, having martyred themselves "for the sake of the team." In between are many reasonable people—some of them thriving, some of them struggling in jobs that are defined mostly by what people want from them while lacking clarity, role models, or sufficient support. We could start with what the team needs, but the goal is to effect change, and you can't give what you don't get; so we start here, with you.

DRIing Your Career

You are responsible for—the DRI (directly responsible individual) of—your career. Who else would it be? Your career is not a place for passivity—this is a place for clarity and ownership of your decisions and the trade-offs associated with them. This advice is given out a lot, but we also have to remember to *take* that advice—especially when things aren't going the way we want them to, for reasons that feel out of our control.

Conflating career growth with promotion is a common failure mode in ICs and managers alike, as promotion is often a political game, out of your control, or is the responsibility of your manager. As a result, defining career growth around promotions is a disempowering and ineffective way to approach your career.

In this section, we'll work through what it means to be the DRI of your career by figuring out what you want from your career and how your current job fits into that (or doesn't), making a plan for your own professional development, and recruiting the resources you need to succeed.

I find the mindset of DRIing your own career incredibly helpful, both as an individual who has always felt like the DRI of my own career and as a manager, because I think it makes the boundaries of what you can and can't do for other people clear.

What does it mean to be the DRI of your career?

- Understanding your career decisions and optimizations
- Setting and executing on career goals
- Becoming more coachable
- Knowing when to move on (and to what)

We'll cover these topics more in depth as we get to them, but first, let's briefly look at what each of these means.

Career Decisions and Optimizations

Our current job is just a moment in the overall arc of our career. As such, we should both understand the moment—this job and what we're getting out of it—and have some idea of what we want the overall arc to become. We will look at what kinds of options you want to be available; what is available to you now, in this moment; and what it looks like to make conscious decisions about what you are optimizing for.

Expect more from your career—and less from your current job. Your life is more than your career, and your career is more than your current job. *The goal is not to optimize for this moment—the current job—but for your overall career.* Sometimes, jobs you don't enjoy still contribute meaningfully to your career trajectory. Sometimes, jobs you enjoy are not developing your skills. Being clear about the distinction makes trade-offs explicit and decisions clearer.

In Chapter 1, we'll cover pitfalls that can undermine your overall market value and the trade-offs you can—consciously—decide to make (or not). *The goal is to think critically about how you want to maintain and build your market value.* If your job doesn't match the market in a way that will eventually make it hard for you to find another one, hopefully your employer is paying a lot of rent (i.e., compensating you generously) because that job is eroding your market value. At times, that might be worthwhile, but often it's not, and people realize that too late.

Setting and Executing on Career Goals

In Chapter 2, we will cover what it looks like to define good career goals, or proximate objectives,[1] and how to coordinate the resources we need to be successful at them.

Professional development plans are often produced during the review cycle and then quickly forgotten until the next one, making them not very useful. However they can be useful, and you'll learn how to define one that works for you. This will help you define your proximate objectives: your next achievable goals that build the kind of career (and life!) you want. The goal is to take ownership of your own development and consciously decide what you have to invest (attention, time, and money) in it and how to maximize that. Your professional development budget is what your employer gives you to invest in your career (take advantage of it!), and what you use it for and decide to invest on top of

---

1 The concept of a "proximate objective" comes from *Good Strategy, Bad Strategy* by Richard Rumelt (Profile Books)—it's the next achievable step on the way toward your goal. We'll be coming back to this idea regularly in different contexts over the course of the book.

that is up to you. Having a sense of what you're working toward both within and outside of your workplace helps you prioritize time (and money).

We'll look at the full spectrum of our support system, what we can look for from whom, and how to build it. The goal is to have a strong network not only in our workplace and our industry but also in the rest of our lives, which will make us more grounded and resilient. Often, who ends up as our manager is outside of our control, and that makes people feel powerless (especially when they expect too much from their job). But we have power over all our other sources of support, which can buffer us against reorganizations and our managers' fallibilities.

Becoming More Coachable

In Chapter 3, we'll cover what it means to be coachable and how to be *more* coachable. This is about being open to change, taking available support, consciously expanding your perspective, and turning feedback into something actionable.

Constructive feedback, given at the right time, can help us grow; however, it can be hard to come by. Having a better relationship with feedback, understanding how to get it and how to make the most of what you obtain, can unlock more opportunities by making you a more rewarding person to help. *The goal is to understand and maximize the contexts in which you can get the most actionable input, and to know when to let things go.*

Feedback can feel very threatening, but you can shift your mindset such that feedback is just information. Sometimes it's information you can use to improve, sometimes it's information about someone else (i.e., how they like to operate or what they value), and sometimes it's information you can use to make informed decisions about where you should be spending 40 hours a week. We'll cover how to get the most out of the people around you, how to make sense of the feedback you do get, and when to put feedback in the bin. *The goal is to improve your relationship with feedback and maximize what you get out of it while minimizing the harm.*

Knowing When to Move on (and to What)

There are signs to pay attention to when it comes to thinking about quitting your current job and how to define an action plan based on where you are now and where you want to go.

Understand the warning signs that indicate when it's time to find a new job that will better support your life, career goals, and overall well-being. *The goal is to think critically about whether to stay or go and to consider what you can try to make your existing situation work.* Sometimes people stay too long in jobs, and "five years" of experience is really more like one year of experience, repeated five times. But sometimes people leave a job without expressing their needs or setting boundaries, resulting in them going on to make the same mistakes in the next job. You need to be able to know if you're going *toward* something or *away* from something. As every horror movie will remind you, the things we run away from catch us up again.

Owning Your DRI Responsibilities

This final section will provide you with a clearer idea of your long-term options, your short-term goals (proximate objectives!), and tactical steps to build up any gaps holding you back (controlling weaknesses)[2]—whether that's your relationship with feedback, your specific skill gaps, or your network.

Your career is a much longer-term play than your current role, but it's easy to get lost in the pressures of right now and neglect the longer-term thinking that will provide compounding gains down the line. This is your opportunity to step back and think critically about what you *genuinely* want (not what you *should* want) and what will really bring you growth, joy, and a career you love.

---

2 The term *controlling weaknesses* is from the book *First, Break All the Rules* by Marcus Buckingham (Gallup Press). Weaknesses are things that everyone has but don't necessarily need to work on because it's more impactful and rewarding to focus on strengths. *Controlling weaknesses* are the weaknesses that hold you back, and there's value in getting good enough at them so that they're no longer a problem.

Career Decisions and Optimizations

Career decisions often feel fraught and scary. Do we say yes to the big project? The new job? What are the implications? What if we fail or leave and regret it?

If we don't have exciting new opportunities to hand, we can often second-guess ourselves. If we enjoy our job and feel comfortable doing it...are we not pushing ourselves enough? Have we stopped learning? Are we...lacking ambition? Should we interview somewhere else just to see what's out there? Start a(nother) new side project? Switch to another team?

The risk of the career decision is that we get it wrong. But the more insidious risk is that we make a nondecision decision; because we are afraid to be wrong, we implicitly decide to keep things as they are, even when that is not working for us.

Contributing significantly to the inner turmoil is how loudly the world tells us what we should want. Years ago, Sheryl Sandberg exhorted women to lean in, and I swear that message is still echoing within me. Every week there are new think pieces warning about "quiet quitting" or why going to the office is critical for getting ahead, and viral social media posts of career and/or productivity advice abound. The cacophony of voices telling us that we are somehow not enough and probably never will be is exhausting.

You will not find any of that here; I don't believe in offering advice[1] on almost anything.[2] How could I—or anyone—possibly know what is right for your career? Only you can really understand your constraints and appetite for risk and

---

1 And yet I still decided to write a book.

2 The three topics I will confidently offer advice on are: (1) don't fracture your shoulder; (2) when your job gets harder, the best thing you can do is make a friend; and (3) skin care.

change, what you do—and don't—enjoy doing, and how your career supports what you want from your *life*.

What you will find is a framework you can use to structure your thinking and make conscious, deliberate decisions about what (and why) you want now and how that fits in with what you want later.

Expecting More from Your Career and Less from Your Job

The first point in the mentality of being the DRI of your career is expecting more from your career and less from your job. Your current work situation—good, bad, or fine—is just a moment in your career. **The goal isn't to optimize for this moment—the current job—but for your overall career.** Your life is more than your career, and your career is more than your current job.

We'll break this into three concepts:

- Planning for opportunity
- The work > the title
- Defining the moment

PLANNING FOR OPPORTUNITY

Some people have a five-year plan. I have optionality.

I've never been able to connect with the idea of being able to plan my life. This is even truer in a relatively new, highly evolving industry. Who's to say that the things I'll be working on five years from now are even viable today? If I plan, I'm necessarily limiting myself to what I can see today, leaving out all the things I'm not yet aware of.

I try to think about things less as "What do I want to do?" and more in terms of *optionality*, by asking the question "What opportunities do I want to be available to me?" The more options something opens up, the more excited I am about it. For instance, investing the time to take coaching training increased options available to me.

However, decisions that cut off options should be made carefully: by the very nature of removing options, you remove trajectories of growth. The more I commit to the management track, the harder it gets to go back to being an individual contributor. When I first shifted into management, I didn't really think through the consequences, but when I shifted from my first to second management role, I thought deeply about whether I wanted to continue on that track, what the alternatives would be, and what it would take to get back on the IC track (a calculation I periodically rerun).

Some decisions cut options off indefinitely, but some just cut options off for a time. When I decided to write a book, I knew I would need stability to carve out time for that and wouldn't be looking to shift jobs or to take on too much more responsibility while the project was underway.[3] The short-term trade-off was worth it to achieve a long-term goal that I hoped would provide more optionality later.

Story

When I dropped out of Harvard Business School (HBS), everyone thought I was making a terrible decision. Many commenters on popular MBA blogs speculated that I had "hit the screen" and failed out!

But I knew that I had to drop out: I'd taken an internship as an entrepreneur-in-residence (EIR) in an incubator startup. The nascent project turned into a real travel company, Lola, where I'd be the first employee and head of product. This presented an amazing opportunity that I would not have been able to experience while still studying at HBS. I was getting to work with cutting-edge technology and interesting interfaces. It was perfect, and it opened up a number of options for my career.

Plus, things have a funny way of working out. Years later, during the early pandemic, I ended up at a transitional moment in my career, and I resumed the program at HBS. Finishing up at HBS gave me the perfect time for exploration and growth before joining boldstart, where I became a partner. It's important to recognize what options can open the best opportunities for you—sometimes they even come back around.

—*Ellen Chisa, Partner, boldstart ventures*

REFLECTION

Think about what options you would like to be available to you down the line. What could you do to make those options more available to you? If you're a planner, try an experiment to reframe your plan in terms of making options available to you. How does this change things?

---

3 Deep-in-finishing-the-book Cate laughs (hysterically) at the past Cate who thought she might pull this one off.

THE WORK > THE TITLE

A job title is a few words. The work is 40 or more hours a week. It makes sense to prioritize accordingly.

A job title isn't a goal. Job titles can be useful, especially for people who get judged on past performance rather than potential—in particular, this occurs with people who are historically marginalized. Once I got a "proper" job title (Director, Native Apps), I started getting much more interesting recruiting messages compared to when my job title was defined in emojis (📱🔨), even if both roles were functionally very similar. I also got a lot more sales emails, so it remains unclear as to whether it was a net win.

The big problem with job titles is that they aren't comparable: they exist in the context of an industry and an organization's structure, values, and modes of operation. As a result, they are often meaningless and are usually far less impactful on your life (and career) than the actual work you do. Over-prioritizing factors relative to their impact is a fast track to making decisions that don't best serve your overall well-being.

Do you want to be a "staff engineer," or do you want technical leadership on complex, interesting projects? Do you want to be a "Vice President of Engineering (VPE)," or do you want a job that combines organizational and technical leadership of a large organization? Using job titles can seem like a useful shorthand, but it's easy to default to chasing status over what will actually make you happy. Be specific about what you want and why you want it. Think critically about the work you actually enjoy and how you add value.

REFLECTION

What do you want to be *doing* next year? Five years from now? What are the activities that make up your day-to-day?

Finally, consider how some job titles can reduce your optionality and make finding your next role more difficult. Even considering how titles can scale across different-sized organizations, once you take a manager title, it becomes harder (although not impossible!) to go back to being an IC. Once you have a VP title, it can become harder to find another job at this level—even if it's just because there are fewer of these roles available. While recruiters often find people based on keyword searches including job title, it's common knowledge that job titles are often meaningless. If your responsibilities and scope don't match your title, this will usually come up very quickly when you are interviewing, and that may count against you.

DEFINING THE MOMENT

This job is just a moment in your career.

Because people tend to define their careers through the lens of their current roles, they attach too much importance to what's currently going on and miss its place in the bigger picture. Whatever is going on right now is just a moment in the broader arc (Figure 1-1). Your career is not defined by this any more than a monthlong adventure is defined by one day within it. Yes, occasionally, in extreme circumstances. But it's very rare.

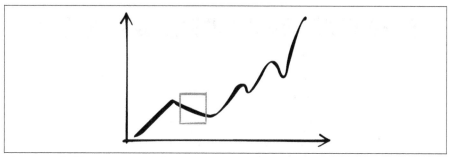

Figure 1-1. The bigger picture is the most important thing; even if you see a small dip in trajectory, the overall trend is moving up

Deciding what this moment is helps you decide what to do with it:

...a moment of opportunity
What potential does this create? What optionality does it facilitate?

...a moment of challenge
This is the power of the stretch assignment—meet the challenge, see what opens up as a result.

...a moment of trauma
The most dangerous moment—is this moment creating something you will carry with you and need to untangle later? Tread carefully.

...a moment of calm

Sometimes we need our jobs to just be fine—not too stressful, not too challenging—to create space for other things in our lives.

Looking at time spent in roles or organizations as moments can help give perspective and make things more endurable. Maybe your work situation is horrible, but you choose to finish the project/organize your finances before you move onto something new—despite how miserable that sounds, there is huge power in choosing to endure something for your own reasons rather than being a victim of circumstance.[4] Maybe you take on a challenge and push yourself because you know it doesn't need to be forever and this opportunity is worth it. Maybe your ambition did not die because of COVID-19; instead, you just needed a moment of calm to survive living through a global pandemic.

REFLECTION

If something feels untenable, how do you step back from it and make it a moment? What needs to change? And what resources are available to you to make that change?

It's critical that we step back to consider what we *genuinely* want—not what we think we *should* want or what we think we *might* want—and how to create and validate those options.

REFLECTION

What are the parts of your job that feel most worthwhile and rewarding?

The trap of the "should" want is why the job-title trap can be so risky: we push ourselves to chase a title without taking time to think about the work we truly enjoy. The trap of the moment is that it destroys our boundaries and puts us in a place of "when our job sucks, our life sucks." A bad relationship with our work is rarely contained to the time we spend doing it—instead, it spills over into

---

4 All credit to my therapist for this revelation!

our evenings and our weekends, through our conversations, and into the energy it drains from us.

Even if you have excellent boundaries, 40 or more hours a week is a lot of time to be unhappy: figuring out the *optionality* you want to be available to your future self helps you define the current direction, understanding the *work* you really enjoy gives you clarity about what you're moving toward (independent of the title associated with it), and defining the *moment* gives a sense of purpose. Understanding these three things can help you redefine how your current role fits into the overall arc of your career—and how that interacts with the kind of life you want.

Story

When I joined my current company, I had a clear five-year plan: work as a resident engineer (construction management) in the field at locations across the US, learn contract management practices, and save enough to travel the world. The goal was to gain the skills and abilities in contractor negotiations and to handle construction management to either build my own home or undertake significant renovation projects. This plan also included a dream of saving up money from my job (and not having to pay rent due to being on the road 100% of the time) so that I could eventually take some time off to go backpacking through Europe and Asia, embracing the adventure that would be youth hostels in shared rooms of up to a dozen strangers, all while making memories through shared experiences.

However, the onset of COVID-19 brought unforeseen changes. Fieldwork was halted, and I found myself climbing the ranks rapidly in a remote setting. When we returned to the office, the landscape of my career and personal aspirations had shifted dramatically. The idea of shared accommodations in hostels, once an exciting prospect, no longer held the same appeal in a post-COVID world. I also was now much further advanced in my role, and as a lead engineer, I only worked out of the office in a city I never intended to live in long term.

This unexpected twist forced me to rethink my original plan. The skills and experiences I gained were invaluable, but the path I had envisioned for myself was no longer viable. I had to adapt, reassess my goals, and develop a new plan that aligned with the changed world and my evolved aspirations.

The experience taught me the importance of flexibility and the ability to pivot when circumstances change. My journey is a testament to the fact that while it's crucial to have a plan, it's equally important to be open to rewriting it in response to life's unpredictability.

—*Alexander Coffman, project engineering lead for the federal government*

Distinguishing What Your Employer Rents Versus What They Buy

I used to work at a company that—even at more than 1,000 people—rejected a lot of traditional wisdom about how tech companies should be run. There was no product management, no CTO, no VPE, no promotion or leveling system. Some people might have run from such a situation, but I did not. I initially joined to run the mobile team (at the time I joined, a disconnected nonteam of around 20 people) because it was a great opportunity for me and a significant step up from what I was doing previously,[5] and it did not require me to relocate to the United States. I figured I would have a very difficult 18 months riding the learning curve and getting things sorted out, but after that I would have a job I really loved.

Well, half of that ended up being true. After the very difficult 18 months, we became a high-performing team of 40–50 people, and others in the organization were impressed by what we had accomplished. However, there was no pay raise or promotion (because those things didn't really exist). Instead, I was given another team to "fix"—which, OK, I took because I saw how running a nonmobile team would provide more optionality in my career, and because I really value being a team player and doing the right thing for the organization.

Fast-forward another 18 months, and it was apparent to me that my job had evolved into a terrible approximation of a VPE job but without the title (thankfully, I didn't even want the title; see "optionality") or compensation (this bit would have been nice). I was tasked with scaling hiring, managing recruiters (which I hated), and trying to provide support to leads in a system that structurally undermined them. I felt like everyone else's problems ended up on my desk. Team not delivering? Fingers were pointed at us for not having staffed it. Team needed to be dismantled? Somehow my job. Teams not onboarding? *We're trying here!*

Meanwhile, I had no power to make any of the structural changes I'd been able to make when I was in more of an engineering director role: setting

---

5 Running a six-person team at a failing startup while being a questionably legal migrant in Colombia.

standards, supporting people to meet them, and leveling them up to support the needs of a growing team. Being disconnected from shipping product made it harder to demonstrate the kind of outcomes (improved effectiveness and delivery) that made the people-centric work feel impactful.

While there were clearly many things I didn't enjoy about this situation, the thing that really pushed me to leave is this: it was clear to me that a situation where I had responsibility but not power, where I owned process but not outcomes, was a situation where I was making myself less employable on the open market. It was also clear that this situation was a product of organizational values, which are not completely intractable but are often the slowest things to change in an organization. Those values were not going to change on a time frame that I could live with. It's one thing to suffer for growth (during the first 18 months of the role), but it's another to suffer while knowing that every month I spent there, I was making it harder for myself to move onto something better elsewhere.

My employer was taking the thing it rented—my expertise—and eroding the market value of it, all while paying significantly below market value for the thing it bought—my time.

Reframing my thinking this way brought so much clarity to this situation, and when I first shared this framework on my blog, this point was the thing that resonated *the most* with people. Again, we're looking at this from a long-term point of view, and what seems like a reasonable bargain in the day-to-day isn't necessarily a good long-term decision—much like eating 20 avocados a day makes it hard to buy property (OK, the avocado thing never happens except in the *New York Times*, but the job thing totally does).

The point of this is it's an opportunity to step back and consider the components of value that you trade with your employer for money (and, where applicable, equity) and consider if that trade is working for you.

Story

My first job out of college did not give me the best start to my career. I was in a city without a big tech presence, did not go to a prestigious school, and lacked confidence, so I took the first offer I got without negotiating. The role was supposed to start in professional services and transition into product engineering. This expectation changed, and I got trapped in a job that did not align with my goals. I stayed there for years longer than I should have. I was learning but not focused on the right things. I was being stretched by the pressure of business trips and billable hours instead of challenging engineering problems.

Eventually, I got up the courage to request a move to engineering. I was not taken seriously until I contributed to a winning hackathon project. After several long months, I was moved into a junior engineering role despite several years at the company and possessing a deep knowledge of the product. I was learning more, but I was underleveled, underpaid, and undervalued. I was worried that staying longer would stagnate my career forever. I eventually left for a job at Google that offered an opportunity to get my career back on track and a life-changing amount of money. Several years and several jobs later (roughly the same amount of time I spent in that first job), I became a principal engineer. I learned my lesson: do not stay at a job where you are not growing toward your goals. Instead, seek out opportunities for yourself.

—*Julie Pagano, software engineer*

REFLECTION

Think about the work you have done for the past 6 or 12 months. How would you talk about it in an interview? Could you authentically and honestly show that it has prepared you for the role you want?

RENT

When you "rent" something out, you retain ownership of it—the renter just gets the use of it. As the long-term owner of the asset, it's on you to be conscious of the value and make sure the renter isn't destroying the long-term value.

Expertise

Expertise is a key example of what your employer rents. Your expertise is *yours*. It's what will help you get your next job and negotiate your salary (oh hai "market value").

This is why there's discussion about "staying current" as a developer—and why managers worry that if they get away from the code, it will be hard to go back to being an IC. Our expertise is only valuable if we can do the work that is needed *today*, so we have to keep developing our skills with a mind to the future. The more Swift replaced Objective-C, the more important it was for iOS (and

MacOS) developers that Swift was part of their day-to-day, because the ability to write Objective-C was clearly becoming less valuable over time. At university they used to warn us, "There are a lot of unemployed COBOL programmers."[6]

This idea is often where the side-project dialogue goes off the rails. Should you need to do side projects to get a job? Are open source contributions a requirement? I personally say no to both of these points. However, being strategic about where you invest resources (time and money) and/or developing a habit of deliberate practice (*https://oreil.ly/uyjTt*) can help you develop your expertise and command a higher rent on the market. The extent of that investment is up to you (we'll talk more about this in Chapter 3).

Developers, if they are lucky, can largely get away without thinking too critically about this because professional development is a core retention strategy of most "good" tech companies. But at smaller companies, at levels above senior developer, or in management, training is often limited and deficient, and if the individual doesn't take control over their own growth, they will hit a ceiling.

As a hiring manager, the biggest mistake I see people making in this realm is having a resume that shows "eight years of experience," but when you dig into the roles, you see the same couple of years of experience again and again. Some people choose the same types of roles deliberately; they specialize their roles in a way that works for them. However, people who don't choose deliberately, or who fall into this trap early, top out and stop getting recruiter pings, and they don't know why—it's because they didn't consider how to grow their expertise as they switched (or didn't switch) jobs.

Brand

There used to be a lot of discussion about "personal branding," and frankly, it was nauseating. But as individuals, we all have a "brand" or a "profile." At a minimum, this is what's on your resume and how you build the narrative of your career. Some companies on your resume will help you get interviews long past when you worked for them. However, some companies or roles are detrimental to that. If you work at a company that's materially contributing to (and profiting from) the collapse of society, that may reflect on you.

---

6 Now it's more like there are a few COBOL programmers, and they make a lot of money, largely from banks or the government. So I guess sometimes it works out!

Another mistake I see often is where people have a job title or custom role that doesn't match the market—and they don't realize it. Companies sometimes try to pay their "rent" in inflated job titles; I'm extremely skeptical about the value these have on the open market, which is another reason why it's important to evaluate the work you'll actually be doing more carefully than the job title when considering new roles. For a good recruiter or hiring manager, seeing a "VP" or a "Director" job title for a candidate who doesn't possess an expected scope of responsibility opens more questions than it answers. It's worth mapping your responsibilities to the expectations of the market periodically and checking that you're still generally employable.[7] Hybrid roles where people perform two different job functions (sometimes called "and roles") are particularly susceptible to not meeting the criteria of either role externally. An "and role" can be a plus when you do one thing particularly well but have additional skills—for example, an engineering manager who does some product work or a designer who also does some CSS. However, "and roles" where people do two or more jobs poorly don't set them up for success in their broader career.

But for those of us who have a bigger profile or some recognition in the community, the concept of brand goes beyond the resume. In a leadership role or developer relations—but to some extent elsewhere—your profile and ability to attract talent are part of your value to a company. It's important to remember that you retain ownership of your profile and that it's for the long term. You don't want to be seen as a shill for a company that later turns out to be problematic.

I saw a tweet (Figure 1-2) that made me think about this; the company thought they were buying this person's profile. But she was clear it wasn't even available for rent.

This subpoint on brand doesn't apply to everyone, but where it does, it's worth considering carefully. The main thing here is boundaries. Think about how you want to use your profile, what options you want to be available to you, and what boundaries you want to set. Make sure you're balancing between the rent and the ongoing market value.

---

7 My suggestion is to take some recruiter calls and do some interviews. Treat it like an information-gathering session and just see what you learn. You can also go to some industry events, seek out some blogs or books, and try to understand the work of other people who have similar job titles to determine where your gaps are.

Figure 1-2. Tweet from @tarah dated September 15, 2021

BUY

When you sell something, you transfer ownership—it's not yours any longer; it belongs to the buyer. When you can't get something back, you need to think critically before you part with it.

There are three major things that your employer buys when you work for them: time, energy, and adherence. We'll talk more about managing time and energy in Section 2, "Self-Management", but here we'll focus on the exchange being made.

Time

Of all the things your employer buys, time is the most obvious. Whatever bargain we've made when it comes to time—an incredibly demanding job or a flexible job—we sell our employer some amount of time. Upholding boundaries around what is work time and what is not affects both our effectiveness and the overall quality of our lives. Most full-time jobs are around 40 hours per week. Some jobs take up more time than that for whatever reason, and they should pay more as a result.

The time aspect almost certainly contributes to systemic inequity; for instance, research shows that senior roles are more likely to be held by men whose wives are less likely to work. Women in these roles have fewer children, and they're more likely to be single.[8]

More senior roles tend to be heavily scheduled and meeting driven. However, at the IC level, people still have to balance their life around their work or, now that we operate more flexibly and more from home, balance their work

---

8 See Boris Groysberg and Robin Abrahams, "Manage Your Work, Manage Your Life" (*https://oreil.ly/ DJBzY*), *Harvard Business Review* (March 2014); Lila MacLellan, "70% of Top Male Earners in the US Have a Spouse Who Stays Home" (*https://oreil.ly/v5CfZ*), Quartz, April 30, 2019; and Maddy Savage, "Why Promoted Women Are More Likely to Divorce" (*https://oreil.ly/xROVn*), BBC, January 22, 2020.

around their life. Based on my experience of working in a distributed context for years, the biggest mistake people make with time is that they don't manage it well. Some people cannot cope with the flexibility; they fail to give themselves the structure they need to be effective. The other failure mode is the opposite: people struggle to disconnect from work, then overwork and burn out.

Energy

Energy is distinct from time, as I think it's what explains the continual gap between the hours that people work and the hours that people think they work (much more).[9] My theory is that when people think about how much they work, they count the energy they spend. When they track the time, it's more concrete. So while a long dinner with friends is definitely not counted as "work time," if you've had a dreadful week and spend half of that dinner venting about how much you hate your job, it counts as energy spent on work, and you think you "worked" more than you did.

It's easy to keep thinking about work even when you've stepped away from the computer, especially the worst parts of your day. It's worth asking yourself the question of whether thinking or talking about work is actually necessary or useful to you and how you can let it go, enjoy your evening or weekend, and pick it up again the next working day. More extreme, dysfunctional environments are an energy drain way beyond the time commitment, and typically way beyond what you're actually compensated for. Thinking about it this way is an encouragement to set boundaries or—if they aren't respected—make changes (like turning the phone off or ultimately looking for another job).

REFLECTION

Try keeping a list of work things you think about outside of work hours for a week or a month—however long is needed to get a good picture. How many of them are related to tasks you are trying to accomplish or problems you are trying to solve? And how much is related to navigating roadblocks, politics, or other dysfunctions?

---

9 Author Laura Vanderkam—my favorite writer on the topic of time—strongly recommends time tracking for this (https://oreil.ly/eCeGl).

Adherence

Adherence is the agreement we make as part of any employment contract. Some jobs don't allow people to do any programming/writing/external speaking, and for some people, this can be a lot to give up. Or maybe you have to live in a certain place, work out of an office, and so on. In distributed contexts, travel used to be one aspect of adherence. When hiring at Automattic, we were super clear: for 49 weeks of the year, you could be wherever you like, but for three weeks of the year, you would be expected to travel. It wasn't for everyone, but I think that's normal and expected; not everything is for everyone.

If your employer is buying adherence, then it needs to be reasonable for you to adhere to it. Perhaps no amount of money would make you go back to an office—I totally get it—but if your employer thinks they have paid for that, you'll have to figure a way out of that conflict that may well involve finding a new place to work. Personally, I can't imagine agreeing not to write and code for fun ever again, but earlier in my career, I made that trade-off for a while.

TRADE-OFFS

When you think about career decisions in terms of renting and buying, you can consider different trade-offs and options that work best for you at any given time. As an employee, it's typical to allocate more time and more energy and, in return, to receive higher levels of investment in expertise than as a freelancer; this helps people grow more and ultimately earn more over time, even if their hourly rate is lower. Sometimes people want to enforce stricter boundaries around time and energy, which can free them up for other things.

I can't emphasize enough that **all choices are valid**. Most of us make different choices at different stages in our career, and people operate from wildly different constraints.[10] My question is: are you making these choices mindfully? And do they work for the life and career you want?

---

10 It's worth noting that systemic inequity drives many constraints—for instance, caregivers have more constraints around time.

REFLECTION

- What is your current employer renting?
- What is your current employer buying?
- What trade-offs are you making?
    - What trade-offs are not worth it?
    - What trade-offs need to have a time limit?

2

Setting and Executing on Career Goals

When we talk about career goals, it's rare that we talk about things that are actually...goals. Often what we call "career goals" would be better termed "external validation" or "career dreams."

External validation is often out of our control. "Getting promoted" is not a goal. It is a desire to be recognized as being at the next level. "Be a VPE" is not a goal for someone who is currently three-plus levels below; it is a dream.

Dreams are often most useful directionally, as some kind of guiding principle. If an early-career engineer dreams of being a Staff+ engineer, that is a guiding principle that can be very useful in early career decisions. In a choice between leading a small team working on a simple application and working as part of a bigger team with more experienced engineers to learn from, the value of technical complexity versus the leadership opportunity will depend on what the person wants long term; it's important not to underestimate the value of access to people to learn from and time to build depth.

To achieve our dreams, we need a strategy. That strategy will probably contain many goals within it, which show incremental progress and milestones to where we ultimately want to go.

Developing a Strategy

For a long time, I dreamed about writing a book. People often asked me if I was going to write a book because I wrote *a lot*. I wrote things internally at my job and externally on my blog, and I had regular articles in online publications like *Quartz*. But while these things improved my *writing*, they did nothing to address the skills I needed to successfully execute the huge multiyear project of getting a book published and to balance that alongside a demanding job. Writing a book was just a dream until I had a *strategy* to write a book:

1. First, I had to develop a core message that was book size in scope. This took *years*. There needed to be an overarching narrative of sufficient scope, combined with depth in all the pieces of it—so I had to not just build out the message but also gain the experience to back it up.

2. Then I had to turn that into a book proposal (i.e., I needed to learn how to write a book proposal) to convince *other* people that I could write a book that people would want to read, which took an embarrassing amount of time (although to be fair, most of that time was spent in productive procrastination).[1]

3. Then came the...actual writing...of the book, and another strategy of how to make the time to do that, and what to say no to in order to preserve it:

   - I tried the conventional wisdom of writing 30 minutes in the morning but quickly realized that didn't work for me and that I needed to carve out bigger chunks of time. As a result, for the best part of a year I organized my time to preserve Saturdays for writing.[2]

   - I said no to many things I would have "normally" done— I did no public speaking, no blogging, and almost no podcasts.

---

1 A more honest way to put this: I was afraid to write a book proposal, so I wrote about two-thirds of a book instead, giving future Cate the gift of a massive edit.

2 Bye bye, social life.

Much advice about writing a book can be summarized as "come up with idea; work on it consistently (usual recommendation: 30 minutes a day beginning at 5 a.m.); (don't expect to) profit; if you feel discouraged, think about holding your published book and how good that will feel"— maybe that advice is high-level correct, but it's not very useful. Trying to make this concrete and actionable for me was a process of trial and error. Although I felt discouraged many times during this process, as I get to the end of it, I see what I learned from each moment of failure.

It's nice to have dreams, to think about how our life will be once we have the job title we think we want or hold in our hands the book we think we want to write. But we don't achieve things by dreaming about them; we achieve them by developing a strategy, identifying our proximate objectives (iterating on them as we learn), and executing on them. This chapter is about how we do that. We'll outline what it looks like to define good career goals, or proximate objectives, and how to coordinate the resources we need to be successful at them.

Owning Your Professional Development

If you're the DRI of your career, obviously that also makes you—not your manager—the DRI of your professional development.

Often when people think about professional development, they think about the professional development budget that is part of a typical compensation package. That budget is the least of it (not that you shouldn't use it—definitely take advantage of it!). Sometimes the idea of professional development expands to some kind of plan you make with your manager, which usually covers the following:

- What
- Why
- Resources
- Support
- Opportunity

An example of a bad development plan might be something like this:

What
> Demonstrate ownership of a complex feature

Why
> To get promoted

Resources
> $TeamLead will do code review

Support
> Manager, team lead

Opportunity
> Feature X in Y stream of work

Why is this bad? Well, it's not developing anything (except maybe the feature you're going to implement). It's a structured way to say "do this and maybe get promoted" that hides the subjectivity and unpredictability: what if $TeamLead says the code review was too much work to review and you shouldn't be promoted? What if $TeamLead quits? What if the stream of work is canceled?

Development plans like this make people consider the exercise of "professional development" worthless—which is fair, because the plans often are. But if we step back from checking boxes to understand the why behind each of these aspects, we might just create something useful. Let's break it down.

WHAT

The "what" is what you want to develop. A good "what" looks like a proximate objective—a next step that seems totally possible to get to from where you currently are.

As such, a *what* is not a task; a *what* is the growth.

Let's say you set a goal to "improve your public speaking." This is a reasonable topic for professional development, but the "what" of the next step depends entirely on where you're at. Do you want to give a talk in front of a hundred people? A thousand people? Do you want to feel comfortable presenting to your team? Each of these goals will yield entirely different plans.

I'd encourage you to not frame these things as tasks that can be checked off. If you present to your team once, and you feel so bad afterward that you vow to never again speak publicly, you could check a box ("presented to team") but not achieve the underlying goal. Similarly, if you deliver a five-minute sponsor presentation that marketing put together for you, whether it's to a hundred people or a thousand, it's likely that won't really move the needle for you either.

A good "what" is specific enough that you know when you've achieved it but broad enough that reaching for it demonstrates a genuine expansion of your capabilities.

For example:

- Drive more consensus on the team around at least one (but ideally two or three) larger-scale technical improvements by producing written documents or presentations that explain the ideas, and gathering constructive feedback.

- Improve project planning by more accurately estimating (not padding) projects and more effectively using spikes to derisk things. Aim to deliver three or more projects within 10% of the original estimate, including one that is estimated to take three-plus months, while reducing the number of surprises and weekends worked on "last minute" issues.

- Scale up people management by identifying and supporting one to two people who can effectively onboard new hires and one to two people who can effectively lead medium-complexity projects.

WHY

The *why* is your underlying motivation, and it's a critical counterpoint to the *what*. When you read it, the *why* should remind you why it's worth the struggle and the work to take on the *what*.

As such, the why needs to be something that genuinely matters to you. Maybe "I need to do this to get promoted" is it, but if you've read *Drive: The Surprising Truth About What Motivates Us* by Daniel Pink (Penguin), you'll know that "if-then" rewards rarely motivate us; instead, we're more likely to be motivated by autonomy, mastery, and purpose.[3]

Let's come back to the goal of being comfortable presenting to your team. What might be a bad *why* here?

"Because everyone else does it."

"Because I got feedback that says I need to do this to get promoted."

---

3 If you're in a hurry, you could also watch the TED talk (*https://oreil.ly/IDDIw*).

Neither of these hits autonomy, mastery, or purpose. Probably—even worse—these reasons touch on the opposite of all those things. Find a *why* that makes it worthwhile to you.

"So that I can explain concepts, answer questions, and get buy-in for my ideas."

Much better.

RESOURCES

This is what you have to facilitate to meet this goal. Resources might include:

- Your professional development budget
- In-house training
- Online training
- Company resources or subscriptions (e.g., Coursera, audiobook library)

Continuing with the public-speaking example, you might have:

- A Toastmasters group
- Online resources
- A speaker coach

It's tempting to skip over this part, but I bet there's more available to you than you think there is. Asking other people for resources they recommend can be a nice (and easy) way to connect with them around your goal as well as create opportunities for further help and support.

SUPPORT

These are the people who can support you in meeting your goal (we'll talk more about building relationships with these people in "Building Your Support System" on page 33 later in this chapter). Hopefully, your manager is on the list. By doing all this work, you've made it easier for your manager to help you, so I'd really encourage you to at least bring this topic to a 1:1.

Here are some other people who can potentially help you:

- Your teammates
- A coach
- A mentor

- Your professional network (even if you just ask for recommendations on social media, you never know where the conversation will go!)
- Your friends and family (even just for emotional support and accountability)

OPPORTUNITY

Opportunity is where you'll accomplish your goal. This can be the most loaded aspect. Sometimes it can seem like there's no opportunity or that the availability of the opportunity depends on our manager giving it to us.

I think of opportunity as being like flowers: sometimes you are given flowers (yay!), but that is not the only way to get them. Sometimes you pick flowers that you (or someone else) planted a long time ago, and if you want to have flowers later, you will need to make sure you plant them early enough that they have time to grow. Opportunities are the same: given, planted, picked.

Given

The most straightforward opportunity. Someone says, "Do you want this?" and you say, "Yes, please."

Planted

The work you do to create opportunity down the line. In the public-speaking example, maybe you submit a bunch of calls for proposals (CFPs). At work, maybe you focus on defining a significant project in the hope that it positions you to lead it. In this space, it's about creating options and accepting that those options may or may not pay off.

Picked

The opportunity you've worked to take advantage of. For example, you get invited to give a talk—because you wrote a bunch of interesting articles. Or you get the opportunity to lead the team—because you did a lot of mentoring/onboarding.

This analogy makes some of the contention around opportunity clear. For example, there's the concept of "exposure" being an opportunity, the snarky response to which is "people die of exposure!"

If you have a bouquet of roses and someone offers you a daisy, feel free to tell them to GTFO. But if you have a handful of daisies and someone offers you a tulip, it may be worth making some space for it (I'm making some assumptions here about the inherent appreciation for specific types of flowers where rose > tulip > daisy). The point is: there's an aspect of reconciliation between what's on offer and what we have, what should be accepted, what could be regifted, and what is suited only for the garbage. Often the mismatch in expectation shows up in what people think they're offering and what they think someone has.

We need to learn to identify what is an opportunity and what's not, and often that starts with asking questions. Some ideas:

- What made you think of me for this opportunity?

- What do you think I can learn from this opportunity?

- What kind of impact will doing this work have?

Once you've evaluated the opportunity on its own merits, you also need to think about your relationship with the person offering it: will they be offended if you turn it down in a way that might hurt you later? Will this affect their perception of you? Separating this out can be helpful for clarifying your thinking:

- Something that is not great on its own merits might be worth doing if it builds a relationship that you value.

- If there's a mismatch in what someone thinks they are offering and your assessment of that offer, then it can be worth clarifying why that is—especially if the offer comes from someone above you in the org chart.

- Sometimes things are sold as "opportunities" that are at least somewhat driven by the need to fill a gap on the team or for the good of the company; it's not actually an opportunity but rather just an attempt to get you to do additional work. The more frank a conversation you can have about what that is, the easier it will be to make a decision.

- Beware that turning down "opportunities" (even if they are not opportunities for you) can be judged negatively—to go back to the metaphor, the person giving you a daisy might not realize you have a bunch of tulips.[4]

---

4 As is so often the case, unfortunately this can be even more true for under-indexed folk.

Note

Sponsorship is the act of advocating for people and giving them opportunities (as opposed to mentorship, which is more about giving advice).

A mismatch of expectations that can happen with respect to sponsorship is that people can think that having a sponsor means the sponsor gives them the opportunity they want. In reality, there's often work to build trust on both sides. Finding a sponsor more often looks like the opportunity to plant some seeds and then pick a resulting opportunity.

A Totally Hypothetical Example

Ana and Bob work in data analysis, putting together reports. They're both managed by Carolina. Bob is more tenured, but Ana is an enthusiastic recent graduate. Carolina asks Bob to take on more responsibility and editorialize the reports. Bob says no, as he and his husband have a new baby. How can he possibly do more work right now? As the most tenured employee, he has all the most complex reports to generate. Carolina doesn't like to push, so she lets it go.

Ana, as she's new, is doing the easiest analyses, and she sees how she can use some simple scripting to make them faster. With all her work done, she asks Carolina if there's anything else she can do. Carolina gives her a rough idea to investigate that she hasn't had time to dig into herself. Ana looks into it, pulls together some analysis, finds some other questions to look into, and puts together some editorial on it. While it's missing some key pieces, Carolina sees a way to get the help she needs while also developing someone who seems to really appreciate it.

A year later, Carolina gets promoted, and Ana takes over her role. Bob is annoyed. Why didn't Carolina tell him what was at stake? He starts looking for a new job.

This scenario could have gone differently. An "ideal" Carolina could have asked Bob what he would need to take on that additional responsibility and coached him through handing off some of *his* work to Ana. This ideal Carolina would have found a way to grow *both* her employees and create more opportunities for everyone, rather than creating a situation where Bob sought a new job while Ana became overwhelmed and immediately thrown into hiring a replacement for Bob.

But we live in reality, and the unfortunate fact of reality is that many managers fall short of the ideal. They are busy operating under their own constraints, lack training, and don't always ask enough questions or offer enough support. This situation could also have played out differently if Bob had asked for what he needed to make the opportunity viable for him within his constraints (by not increasing his working hours). Being the DRI of our professional development also means being the DRI of available opportunity, and unfortunately, that can mean we need to learn to distinguish between opportunity and additional work. The first step of that is thinking critically and trying to have conversations that will get us the information we need to make a considered decision.

PUTTING IT TOGETHER

Try it out yourself: put together a professional development plan for a proximate objective (or two, or three). Find someone—your manager, a coach, a trusted teammate—to talk it through with, and encourage them to challenge your thinking, question your assumptions, and suggest additional resources or opportunities (including those you want to start planting).

You wouldn't start a large-scale development project without some kind of plan (or would you...?), so why wouldn't you do at least as much for your career growth? Thinking critically and deliberately about what you want and how to get it can help you be more effective with your time and resources. The less time you think you have for professional development, the more reason you have to plan.

As you go, you will need to iterate. If you try to improve presenting to the team and realize that improving your writing and starting with a document rather than a slide deck works better for you, great! Change your approach. You may end up wanting to improve your presentation skills later, but for now, if you're getting the outcomes you want and growing, that's fantastic. If you try to improve your project estimation, but the first project you try your ideas on blows up in scope, think about what you can learn from that and come back again with new ideas for the next one. If you try to scale delegating one thing to one person and it doesn't work out, think about how you can change your approach (and/or who you are delegating to) and try again.

Even if you're lucky enough to have a good manager who thinks about developing you, they're constrained by the needs and options of your current environment, whereas your career expands way beyond that. Don't expect your manager to be perfect—they're busy, stressed, overwhelmed, and not getting the support *they* need. You'll get more useful help if you make it easier for them to help you. Come to them with the ideas you have, what you've tried, and how it's gone and ask for their input.

Building Your Support System

We talked about how you want to grow, and it's clear that some help can often go a long way in accelerating that. As the DRI of your career, you need to build your support system. Not only can this help you grow, but it can give you a more constant basis for the support you need even as things change around you (or don't go the way you want).

I firmly believe that we should get different things from different people—from our managers, from our peers, from our friends, from a coach. Part of managing up (managing your manager) is knowing what your manager is good at and what they aren't good at, and making sure that regardless of that, your own needs are met. Often, who our manager is is out of our control, and that can make us feel powerless (especially when we are expecting more from our job rather than our career, as we discussed in "Expecting More from Your Career and Less from Your Job" on page 8). But we have power over all our other sources of support, and this can buffer us against reorganizations and our managers' fallibilities.

There are many things that hold newer managers back. One is expecting to be able to give their direct reports everything they need. However good the intentions, this isn't possible.

I'll share an experience I had early in my career. I felt the need to build a network of other women to help navigate an environment where there were very few other women.[5] My (male) manager was offended that I felt the need for this kind of connection; he thought he should be my main port of call rather than others with shared experiences and knowledge. I took his feelings under advisement[6] and kept building my own network of people, both internally and externally to that organization. I left that manager behind a long time ago, but many of the relationships I built during that period are with people I still talk to and value deeply to this day. This community I had surrounded myself with

---

5 There were no official numbers, but we calculated there were between 5% and 10% women engineers in that office, considerably below the standard (low bar) of the industry in general. Because laughing is better than crying, we would joke that we had broken the standard "Dave:woman" ratio and moved on to more obscure names.

6 This is British for "I ignored them."

has proven to be valuable for their insights, shared experiences, and industry knowledge—it has gone far beyond navigating one company. It's a reminder that managers are dictated by the organization, but the network we *build* ourselves— for better or worse, that's a product of the decisions *we* make, about who we connect with and how we maintain those connections. As a manager, I can't do the work of building those networks and connections for my directs, but I can encourage them to do it for themselves.

It's understandable that managers want to feel useful, but it's not a failure if the people you manage get some of their needs met elsewhere, especially if this includes needs that aren't realistic for the manager to meet. This mindset of needing to meet all the needs of your direct reports limits the responsibility of the manager to what they can know enough about—like managing more junior people on a less challenging thing—and supports a small monoculture that doesn't encourage people to step outside the bubble and fulfill their potential. As a manager, I've become increasingly clear with my direct reports about what I can meaningfully help them with and what they could better find elsewhere (and then I help them find that).

Whether you have a great manager or a terrible one or something in between, they are your manager for just a moment in your career. It's worth trying to maximize that relationship (and we'll talk about how to do that), but looking outside that relationship is key to your growth and for weathering change—whether that change is a reorg, your manager's departure, a promotion, or your moving on.

Ideally, you have a broad support network. So let's talk about key people in your network, why they are important, and the impact they have.

MANAGER

Your manager is usually the person with the most impact on your day-to-day work. They're the person who will advocate for you (or not) during performance-review season, and they're often best positioned to give you feedback and help you grow.

Determining your approach

Your manager is a human being, with good days, bad days, other responsibilities, and other things on their mind. Many individuals are resistant to "managing up," as though it is a completely unreasonable thing to do. This holds them back. Managing up is your part of making that relationship successful and productive. Is it all on you? Obviously not—your manager has more power. Can you meaningfully contribute? Yes. Can this help you be happier and more effective? Absolutely.

When thinking about how to manage up, consider the following questions:

- How does your manager like to communicate?
- How do they tend to give feedback?
- What are their strengths?
- What do they avoid?

Finally, consider: what do they notice, and what do you need to tell them? Often managers who pay close attention to their own teams and proactively offer support get frustrated that *their* manager doesn't notice the same things—whether that's workload, or getting credit for something they did, or a problem peer. Before getting too annoyed, try telling your manager the thing you would have liked them to notice and asking for the help you would like them to give.

Pitfalls

The most obvious pitfall is having a bad manager, which is...not uncommon, but it's worth doing some validation before you write your manager off. If your manager makes you feel less capable—whether it's through the way they allocate or review work, communicate with you, or operate more generally—then that can be a strong sign that it's time to find a way for them not to be your manager anymore. If they just aren't meeting your needs in some way, then managing up is your first step in trying to get them to. For instance, if you learn that they know about the company priorities but don't proactively communicate them to the team, you can try asking them questions about priorities to draw the information out of them earlier. If they tend to criticize details without looking at the structure first, you can try ways to elicit structural feedback prior to getting the detailed feedback.

The less obvious pitfall is that your attempts at managing up seem insincere or self-serving. If your manager is adequate and reasonable, being considerate and consistent will go a long way.

Top tip

Find out your manager's communication preferences and see what you can easily accommodate.

Key question

"How can I help you?" shows your willingness to pick up things for the team and can help you discover what you didn't know you didn't know.

Story

I grew up and went to university in Uruguay; however, three years after graduating, I moved to Ireland to take a role at a financial services company. This felt like a huge leap: I was still new in my career and needed to work (and live!) in English. Tara, my manager, played a crucial role in helping me adapt and thrive in this new environment. From day one, she offered support and provided timely, actionable feedback. Tara not only introduced me to key individuals beneficial to my career but also showcased my contributions company-wide when appropriate. As I got up to speed, she delegated responsibilities like appointing me as the liaison between the Dublin and Singapore offices.

Her guidance extended beyond daily tasks, as she advised me on project selection to maximize learning opportunities and advance my career. As an example, Tara highlighted manual tasks within the team that could be automated, inspiring the development of a dashboard system that I designed and did the initial implementation for. This project gained momentum with the collaboration of other team members and became indispensable to several other teams in the company.

Tara served as a role model during a formative stage in my career. Her efforts not only paved the way for a swift promotion but also ensured that I felt consistently productive, engaged in continuous learning, and genuinely enjoyed my work throughout my time at the company.

—Eli Budelli, former head of apps, Automattic (turned vintner)

SKIP-LEVEL MANAGER

Your skip-level manager (your manager's manager) will have more influence across the organization and is often a more helpful person for growth opportunities than your manager is, as they have more teams and a broader sphere of influence.

Determining your approach

If skip 1:1s are a normal occurrence at your job, make sure you take advantage of them. If not, consider how you can ask for them or take advantage of things like office hours. Make sure you prepare and use the time well.

Pitfalls

Your skip level is your manager's manager and, as such, probably doesn't have granular information and will (hopefully) avoid undermining your manager in their interactions with you. Avoid undermining your manager, such as asking questions best asked to your manager, because your skip level will most likely defer those questions to your manager anyway and you won't learn very much. Focus on what they can best help you with (given their scope, skills, and experience).

Top tip

Focus on the bigger picture. Ask them how the team fits in with organizational or company goals and what challenges or changes they see coming.

Key question

"Do you have any advice for me?"[7] You'll likely get more insight than asking for feedback.

PEERS

Your peers are the people doing the kind of work closest to what you're doing—whether you're an IC, manager, or director—and it's likely you face similar challenges. The more you move up the org chart, the more alignment (or lack of) with your peers affects your teams. If you're lucky, your manager forms a team of your peers and builds that "first team" mindset, and you get that peer support as part of that team.[8] As you become more senior, though, you may have to make more effort to seek out your peers and work for peer support more; think

---

7 This incredibly useful insight comes from Claire Lew (*https://oreil.ly/JLcN1*).

8 The "first team mindset" comes from Patrick Lencioni, author of *The Five Dysfunctions of a Team* (Wiley), and is about being more accountable to your peers (the first team) than the team you manage.

through what teams you depend on and what teams depend on you, and set up regular communication channels.

Determining your approach

This is hugely contextual, but first figure out who your peers are. If you're the only staff engineer on a team, for example, you may need to look to another team to find someone else in a similar role. Ideally, start with some kind of commonality or shared work or goal and go from there.

Pitfalls

A common theme in organizational dysfunction is that people—especially at higher levels—are pitted against one another. No one wins in a zero-sum competition, but some people lose more than others. The most useful advice I've found on this topic comes from Adam Grant's book *Give and Take: Why Helping Others Drives Our Success* (Penguin), which is well worth reading in its entirety, especially if this situation resonates with you.[9] But to summarize, if you're inclined to "give" to others but find yourself being taken advantage of, look at how you can be more strategic and "match" instead—that is, don't uncritically give to someone who doesn't give back, ask for the things that will help you, and make it clear it's a two-way endeavor. Another tip: if you know someone doesn't give you credit for what you do, look to mitigate that by (1) putting things in writing where possible and (2) giving your boss an FYI of your actions (working this into the format of your 1:1 agenda is helpful), so they are aware of what's happening.

> #### Top tip
> Demonstrate you're a supportive peer by sending colleagues genuine compliments about their achievements or passing on great things you hear about them at work.

> #### Key question
> "How can we help each other?"

MENTORS

I'm generally pretty skeptical about mentorship and largely view formal mentorship programs as Ponzi schemes designed to distract under-indexed folk from

---

9 I credit this book for taking me from existential crisis to manageable problems with concrete solutions in the space of a week.

genuine advancement.[10] But a less cynical attitude is to see mentors as anyone you can learn from who will take the time to help you.

We all need help sometimes, whether it's providing an overview of a complicated debugging tool, giving a thorough code review, or being able to talk through a problem together.

Determining your approach

Ask questions and ask for help. There are so many people out there who truly love to help once you give them the opportunity.

Think about the people you work with regularly: what do they do particularly well that you could learn from? Consider asking for suggestions or advice.

When you see something that someone does well, consider what you can learn from it. If you find a great technical communicator, for example, you may learn not just about the topics they talk about but also how they do so in an engaging way. If you've already consumed content and reflected on it, you'll have better questions to ask them and it will seem more worth their time.

If you find a great peer, consider what you can learn from each other and offer in return. For instance, when my boss hired a VPE, I offered myself up to support his onboarding—and in return got to learn from his wealth of experience. Win-win!

Pitfalls

I see people falling into two main pitfalls:

- They ask for the wrong help.
- They aren't respectful of someone's time.

You will get your best mentoring from someone when you make it the best experience for them: ask them for help in a way and on a topic that they enjoy, and show that you're a good use of their time (we'll talk about coachability next!).

So if someone hates pair programming but loves code review, ask them for a thorough and incremental code review. If someone hates writing, requesting a written report is probably not the best way to get their feedback on something, so ask them if you can schedule a call instead. Paying a little attention to what

---

10 This might seem like too spicy a take, but the data backs it up: "Having a mentor increased the likelihood of promotion two years later for men, but had no effect on promotion for women" (Herminia Ibarra, "A Lack of Sponsorship Is Keeping Women from Advancing into Leadership" (*https://oreil.ly/YBYao*), *Harvard Business Review* (August 19, 2019)).

someone spends time on can tell you a lot about them, but you can also ask people what they enjoy and how they most like to help people!

People often look for the most experienced or senior person they can find to mentor them, but those people are often too far removed to be too helpful. A tenured VPE used to managing directors isn't the best person to mentor someone on management 101, and they're unlikely to see it as a good use of their time. Someone one or two years ahead of you likely remembers more of the things you're running into and has more patience for them.

Top tip

When thinking about finding someone to mentor you on a topic, take the time to think about who, what, and why. See if you can match someone's superpowers to your needs.

Key question

"Do you have any suggestions for resources that I could use to learn and improve?"

SPONSORS

A sponsor helps connect you to opportunity and advocates for you when you're not there. Sponsorship can be a huge accelerator for your career.

Determining your approach

The first thing to understand about sponsorship is that you need to earn it. A sponsor uses their reputation to advocate for you, and it's reasonable that before they do that, they want to know that you're worth it and they can trust you. This brings us to the second thing to understand about sponsorship: sponsorship is about what someone says when you aren't in the room, which can make it hard to know when it is happening—the times when someone advocates for you but doesn't get the outcome and the times when no one advocates for you look the same from the outside.

The most obvious sponsors are your manager and skip-level manager, but it's worthwhile to look more broadly: someone who is technically a peer on the org chart but has more experience or organizational clout can be a great sponsor. For instance, if you're a midlevel engineer, your tech lead may be a good sponsor; even if you have the same manager, the tech lead has more organizational power than you do and more access to opportunities to give you. If you're already in a leadership role, you could look to another director or a VP leading a larger or more complex part of the organization.

First, think about what you can offer them and opportunities to help in return for their sponsorship. I once built a relationship with a peer by helping him with a bunch of hiring work; he had more clout in the organization, but I had more experience hiring and made myself useful, which built trust. Then, think about what you could use their help with...and note that the first request for help should not be a request like "advocate for me" but rather more along the lines of "I'm trying to get to $outcome and navigating $problem, and I could use your advice or influence." Focus on the outcomes, not the personal request.

If you've worked through the development plan earlier and have something you're working on, great! Bring them the *what* of your goal and ask them for advice, but don't neglect to mention the *why*—then they will know what you're looking for. Make sure you come back and update them on any progress you've made.

Sometimes we talk about sponsorship like it's an instant fix, and yes, sometimes we get lucky: a new leader sees our potential, gets us some opportunities, and advocates for us, and that can change a situation quite quickly. But more often, sponsorship is a slower burn. We need to find the opportunities, build the trust, and wait for the moments of advocacy to work out and the impact of the advocacy to come through. Don't be discouraged if your concrete goal doesn't come through, especially if you're making progress against it—if you don't get promoted but you *do* get some more impactful assignments, that can be a short-term win. Remember, it's also a win if you don't get promoted but you *do* understand why and what you need to do about it, and you have the opportunity to do those things—even if it doesn't feel like it at the time.

Pitfalls

A sponsor doesn't have to do that much to meaningfully contribute to your development. They don't have to be a good manager or mentor, or really support you in any way beyond helping you get an opportunity. The biggest pitfall is that people miss the opportunities available to them because they don't look as tidy or straightforward as they'd like.

Top tip

Remember that when someone sponsors you, they lend you their reputation. Take that seriously.

Key question

"Why do you think this would be a good opportunity for me?"

COACH

Coaching is one of the very few relationships any of us can have where it's just about helping us be our best selves and live our best lives. No matter how supportive a manager, our friends, and the like are, they have a vested interest in our actions and decisions. A coach exists separately from everything else and focuses on supporting the coachee's agenda.

Determining your approach

A recommendation is a great way to start finding a coach, and it's worth connecting with a few people to see who you feel the best fit with: a good coach will challenge your thinking and push you to be better without making you feel judged.

At the start of a coaching relationship, you will typically "design the alliance" (*https://oreil.ly/7qkiU*) and talk about how you want to work together; you'll also often set some kind of focus and intention for your sessions. Be intentional about what you want to get out of coaching and how to make the most of the time; try to show up present and willing to be open.[11]

Pitfalls

Most of the work of coaching happens outside of coaching. You have to be willing to really show up for it—there's very little a coach can do if you don't open yourself up to them and take the actions you agree to. This is why finding someone you really connect with is so important; it will help you be more honest with them.

Top tip

Think about how you can best set yourself up for success. For instance, if you put your coaching call in the middle of a bunch of stressful or tactical meetings, you'll probably have a harder time switching to the bigger picture. If you need more structure and accountability, you need to make sure that you and your coach create it together.

Key question

"What do I want to get out of this?"

---

11 Incredibly tactical, but I realized that being based in the EU, I also needed a coach in the EU so that I could have our call during a more peaceful time of day when it was easier for me to be present and I was more able to see the biggest challenge I was facing, rather than whatever I was dealing with prior to the call.

PROFESSIONAL NETWORK

Professional networks are a huge source of opportunity and learning, and they can be critical for checking the bubble of our current work environments and maintaining our perspective on the broader industry.

Determining your approach

Many people hate the concept of "networking," and really, I feel you.[12] But seriously, there's a quote about exercise—"The best exercise is the exercise you enjoy"—and I think the same applies here. If you don't enjoy attending random events, don't do it. If you don't enjoy writing blog posts, why bother? What do you enjoy that helps you connect with people? Invest in that. Some people enjoy speaking (whether on stage or on podcasts, or both), some enjoy writing, and others enjoy curating and nurturing 1:1 relationships.

Pitfalls

The pitfall of networking is engaging with people only when you want something. It's important to find ways to continuously invest a small amount and have balanced interactions; that consistency in balance is so helpful for building meaningful, long-term relationships.

Personally, it always feels a bit icky when I leave an interaction realizing that I heard from a person only because they wanted something and that I'll only hear from them again if they want something else. For me, the ambient awareness of social media has always made staying in touch with people seem easier, and it's easier for me to engage by responding rather than needing to remember to reach out.

Top tip

Notice people's achievements and congratulate them, or reach out whenever you find something particularly helpful. Even better, share your key takeaway. "I loved your blog post" is great. "I loved your blog post, the point about X was so helpful, and as a result, I made Y change with Z impact" is even better—and will make you more memorable.

Key question

"Is there anything I can do to help?"

---

12 It's possible that I got into public speaking in large part to avoid having to initiate conversation at events.

WORK BFFS

One of the few pieces of advice I feel good about giving to everyone is this: when your job gets harder, the best thing you can do is make a friend.

This is also backed by the data. Gallup research (*https://oreil.ly/7xmBT*) has consistently shown that having a best friend at work leads to better performance—perhaps related to the findings that people who have a best friend at work are more likely to be engaged in their job, more likely to take risks that spark innovation, and less likely to be actively looking for job opportunities elsewhere.

One of the hardest things about management can be the emotional labor of supporting other people all day. We're managing down and supporting our direct reports, and we're managing up with our own bosses because we know they have tough jobs and we don't want to be one of their problems. But it's easy to feel that no one is looking out for us.

A work BFF can be there for you on days when it feels like no one else is—or can be.[13]

One of the key benefits of work friends is "if you have friends in the company, it's far easier to ask for help without fearing you'll be judged a poor performer. In addition, having friends in the company, especially if they work in other departments, gives you access to information through informal networks you might not otherwise get."[14]

Determining your approach

Finding and building your work BFF relationship takes time, just like any friendship. See who you really connect with and try to spend more time with them. See what happens! And if you're lucky enough to already have a work BFF, make sure you tell them the difference they make to you. The feeling is almost undoubtedly going to be mutual.

---

13 Read about my work BFF Eli (*https://oreil.ly/xrMep*).

14 David Burkus, "Work Friends Make Us More Productive (Except When They Stress Us Out)" (*https://oreil.ly/s-PVI*), *Harvard Business Review* (May 26, 2017).

Pitfalls

It's also worth being aware of how friendship can come across as cliquey, so look for ways to welcome other people into your friendship, and make sure you don't monopolize one another at work events.

Top tip

Work BFFs often emerge from good peer relationships.

Key question

"Do you want to get coffee?"

3

Embracing Growth

When I think about growth, I think about drowning rats and boiling frogs because this is what growth often feels like—or looks like—especially when it's hard and not going particularly well.[1] The rat-drowning school of growth is to throw someone in the deep end and let them figure it out.[2] If they do, they'll become good swimmers. The frog-boiling school of growth is to gradually turn the heat up so the frog doesn't notice, and if they survive, they'll be heat resistant...

This metaphor sounds violent and less than ideal. But what I like about it is that hard periods of growth *feel* difficult. These metaphors capture the feeling of "Can I do this?" and "Can I sustain this?"

For all I like to write about—and practice—what good management looks like, the reality is that many managers aren't that good, and even good managers, operating under their own set of pressures, aren't good for everyone all of the time. Very few people get everything they need, when they need it, to be successful. Everyone has their moments where they're struggling. If we're lucky, these are the moments where we learn most of all. If we're not, they're moments where the best thing we can say about them is "I survived."

The question is: how do we make it more likely that we learn and grow as much as possible? What resources can we draw on?

---

1 Don't worry, no rats, frogs, or software engineers were harmed in the creation of this book.

2 When I was training to be a ski instructor, one day I got to talking with a group of guys in the gondola. They had decided to take their friend—who had barely skied before—up to the top of the hill, saying, "If he makes it to the bottom, he'll be a good skier." This is a story of the first time anyone wanted to pay me for teaching them to ski, or maybe it was for getting him down the hill in one piece.

A fantastic book on getting the most out of feedback is *Thanks for The Feedback* by Douglas Stone and Sheila Heen (Penguin). One concept within it can completely change the way we look at feedback and foster a growth mindset. The idea is that there is a difference between the first score—the feedback—and the second score—what we do with that feedback. Ultimately, the second score is the more important one: the more we do with the feedback we get, the more we can grow. This section is all about maximizing your second score.

The Gap Between Capability and Requirements

Professionally, whether we're growing or stagnating is the gap between our current capability and what's required of us (Figure 3-1).

Figure 3-1. *The rat on the left is drowning in its bucket and has no way to get out of it; the rat on the right is too big for the bucket and doesn't have room to move*

Note

Capability less than requirements = hard growth (or failure)

Capability greater than requirements = stagnation

Requirements a bit more than capability but with help = sweet spot

It can be hard to find the sweet spot of growth, let alone sustain it. So another option can be to go between periods of hard growth and recharging periods closer to our comfort zone. In a hard-growth period—like when trying to launch a new product or coming to grips with a new and difficult role—we might work a lot of hours and/or feel very stressed. In the recharging period, we work normal hours, pass up or defer extra opportunities, and/or take more time off and focus on recovering our equilibrium.[3]

---

3 I know this will sound horrendous to some people, but there are people who enjoy working short-term stints fixing broken organizations, followed by extended time off. They find the crisis interesting, and knowing they have a meaningful break to look forward to afterward helps them focus and get through it.

Something to remember is that we don't necessarily see real growth even month by month; often, it is more year by year. Also, we might be growing in different ways, such as increasing depth rather than breadth. We don't always see growth when it is happening (both in ourselves and in other people) but only when we look back on what we are capable of today versus what we managed to do before.

Note

Growing through necessity = **stretching**: adding necessary capability to succeed at what we're currently doing.
Growing from our comfort zone = **development**: growing in the ways that interest us and laying groundwork for what's next, such as through courses or side projects.

There's a great book by Marcus Buckingham, *First, Break All the Rules* (Gallup).[4] At the core of it is the relationship between strengths and controlling weaknesses. The reality is that everyone has weaknesses, and often those weaknesses are the flip side of people's strengths. They're only a problem if they are limiting factors for someone's growth, a "controlling weakness." The thesis of the book is that **great managers focus on developing strengths**, and when they work on weaknesses, it's to build them up so they're no longer controlling.

It's ideal if you have a manager who focuses on developing your strengths, but *you* can also focus on developing your strengths, and your manager is not the only person who can help you. If we come back to the idea that the sweet spot of growth is being out of your comfort zone while having help and support available to you, it's worth thinking about what kinds of help may be available to you; at different times, different kinds of help might be more valuable (or available).

In terms of what help is available, we often talk about mentoring, coaching, and management like they're interchangeable. They're not. **Good managers use the right tool at the right time.**

A mentor can help someone learn a skill. This is great and useful, but it's important to map the mentoring to the level someone is at. If someone is struggling in the ocean (an open, chaotic environment) and the mentoring is giving advice on how to swim in a swimming pool (a small, controlled environment), then it may not be at all what they need.

---

4 This is one of my absolute favorite management books and usually one of the first ones I recommend to new managers. For a more individual-focused take also based on the CliftonStrengths assessment, there is *Now, Discover Your Strengths*, by Marcus Buckingham and Donald O. Clifton (Gallup).

Coaching is about empowering the individual to chart their own path. A great coach will help you help yourself. In our ocean scenario, the coach will be asking: What can you see? What can you find? What can you make? Coaching is helping the person realize they can get to a small rock, build a makeshift boat, and sail away.

Active help is practical intervention. It may not be helpful (or it may even be harmful) to coach someone on something that's within your power to fix. Effective managers know when to coach, when to mentor, and when to hand the person a life raft.

It's also important to understand the limits. If someone is in a puddle but thinks they're drowning, that's more of a topic for therapy.

When we understand the components, it's easier to understand what kind of help we need ourselves. In a given situation, what do we need? Is it a coach? Or a mentor? Or practical help? Some combination of all of these? Once we are clearer about what it is that we would benefit from, it's easier to identify where we might get it—and ask for it. It's also easier to see what we *don't* need more of. If you already have a mentor for a topic, it's unlikely you need three more, so if you're bringing the same questions to multiple people, think about why you're not able to move forward with the answers you're getting.

Something that surprised me when I was going through coaching training was how many people seem to get into coaching as a way to give advice. Also surprising, once I started actually coaching people, was how many of them *wanted* advice. When people are struggling with something, sometimes it seems like the answer is for someone to tell them what to do. But often the case is that people *know* what to do—they are just struggling to do it. The reasons for that vary, and the core of effective coaching is often getting into *why* someone is struggling to do the thing they *know* they need to do and finding ways to change that. It can seem more *efficient* to be told what to do, but it's rarely as *effective* as figuring it out for yourself.

The co-active model of coaching says: **the coach holds the client to be naturally capable, resourceful, and whole.**[5]

I can't tell you what to do—you know better.

I can't help—I'm not there.

---

5 Co-Active Training Institute, "What Is Co-Active?" (*https://oreil.ly/UBBSK*).

At the start of any coaching relationship, the coach and client talk about "designing the alliance"—setting the parameters of the relationship. Designing the alliance is a really powerful concept in coaching that I personally use to define a lot of my professional relationships. It's the agreement for how you plan to operate together.

The first step of being more coachable is being open to the process. Someone is trying to help you figure out how to be your best self. It's not about *helping* you but about *growing your capability*. Naturally, this can be inherently uncomfortable. Sometimes the people who are the best at coaching others are the hardest to coach because they understand the process well enough to evade it.[6]

In terms of approaching growth, coachability is made up of two factors: receptiveness to feedback and how highly actionable you are (what you do with that feedback).

What makes someone highly receptive to feedback? We can easily identify people who aren't receptive to feedback. They:

- Respond defensively
- Blame others and are reluctant to take any responsibility (everything is someone else's fault)
- Discount the value of the input or the person giving it

In contrast, people with high receptiveness to feedback:

- Listen and work through what the feedback is
- Are self-aware and prepared—they can fit the feedback into their mental model and adjust it accordingly
- Believe they can learn from everyone

Similarly, we know exactly what people not acting on feedback looks like. People who aren't actionable in response to feedback:

- Make minimal changes or adjustments
- Make any changes they do make slowly
- Are very literal (apply the same feedback to other situations without nuance)

---

6 The best book I've read about this is Lori Gottlieb's *Maybe You Should Talk to Someone* (Houghton Mifflin Harcourt)—it's about doing this evasion in therapy, but similar things apply.

But what does highly actionable look like? People who are highly actionable in response to feedback:

- Experiment, iterate, and change behavior (both in response to explicit feedback and implicit feedback and because they're continually looking to try things and improve)
- Return to those who are interested with things they're trying or have tried, and solicit input on how it's working
- Seek out other perspectives and information to learn more

You can think of these two dimensions as a quadrant (Figure 3-2).

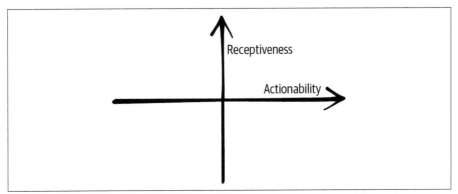

Figure 3-2. Diagram showing different quadrants where the axes are actionability and receptiveness; where your coachability falls will help you determine what you need to focus on to make more progress

It's worth noting that we *should* be in different quadrants with different people: high-trust-high-respect relationships will be further up and to the right, and low-trust-low-respect relationships will be down on the left. This is understandable and, to a certain extent, healthy. While we can always learn from feedback, even feedback given in bad faith or with an agenda, we need to consider some input (from low-trust sources) more carefully than we do input from people who we know care about us and want us to be successful.

Now, let's talk about each quadrant of coachability.

HIGH ACTIONABILITY WITH HIGH RECEPTIVENESS

In this quadrant, everything feels clear. It's easy to fit feedback into your mental model, and adjustments feel natural and build on what you're currently working on. This comes from having a good idea about what's happening and how you think you can improve.

This might seem like the ideal, and on some level it is, but taken too far, being both highly responsive and highly actionable can lead to over-indexing on what everyone else is thinking instead of forming your own opinion, which can make you too reactive and appear a bit chaotic or inauthentic.

For example, let's say you're writing a technical design and want to make sure you're open to feedback. You get a variety of feedback: some nitpicky comments and some more structural. You want to make sure you focus your energy on the structural comments that have the most value: the people with the most context or experience or those who are closest to the problem. If you try to take the same mindset with everyone, it'll be easy to get distracted by the nitpicks and get lost trying to make everyone happy. To manage this, you might share the document first with a couple of people you know you can really trust and whose feedback you want to take; then, once you've incorporated their point of view, you can open it up to a wider audience.

REFLECTION

Think about the relationships and situations where feedback is most clear and actionable to you. Why is that?

HIGH ACTIONABILITY WITH LOW RECEPTIVENESS

This is the quadrant of hard but rewarding work. There's a higher bar for taking input, but the input that is taken is acted on and (usually) results in meaningful change.

In this quadrant, it can feel like the feedback doesn't fit into your mental model, often your self-identity. If it did, you'd act on it, but it conflicts with the other information you have. The first step to moving forward is to reconcile that conflict.

For example, imagine you're a manager who prides themselves on empowering the team, and you get anonymous feedback that you are micromanage-y. You're committed to being a great manager for the team and don't know where this critique comes from—you pride yourself on making sure people have clarity! You hate to get in the details! You are too *busy* with important strategic work to be a *micromanager*!

The first step is to acknowledge that there are reasons why this feedback may not be valid: if you had to put someone through a performance improvement plan and they complain that you're a micromanager, there may not be much to learn from that. Then, unless you have solid evidence this is the case, it's worth getting curious about the feedback.

Maybe there's a subproject where you were more in the weeds than usual because it was risky and really important to *your* manager. Could the person on that project feel micromanaged? Do they have the context you do, or did you shield them from that in a well-intentioned way (you don't want them to worry!), only to inexplicably (to them) get all up in the details?

REFLECTION

Think about some feedback you got that you couldn't make sense of. What does it tell you about the experience of the person who gave it to you? Try and step into the shoes of various people and consider what would make that rational feedback for them to give you.

LOW ACTIONABILITY WITH HIGH RECEPTIVENESS

This is the quadrant of unfounded optimism. Feedback conversations that go great, followed by little to no change.

In this quadrant, you feel a bit stuck. The feedback you're getting makes sense and feels clear, but...you can't act on it for whatever reason. The first step is figuring out what's stopping you.

Let's return to our tech design example. You're designing a system for video playback. You get some feedback and have a great conversation with a key stakeholder, then return to the problem and realize: they are optimizing for playing long videos with no disruption, but you need to optimize for near-instant loading time of quick video clips. How do you reconcile these competing objectives? It's possible you can't (or at least not in the time frame you have), and you'll have to go back to the beginning and clarify the constraints of what you're trying to build, including what the service can support—and what it cannot. Make sure you come back to the person who helped you realize this and let them know, though! Otherwise, they will feel like they wasted their time on you.

REFLECTION

Think about some feedback you wanted to take but couldn't. What got in your way? How did you address it? Did you follow up?

LOW ACTIONABILITY WITH LOW RECEPTIVENESS

This is the quadrant to avoid; it's a poor use of everyone's time.

This quadrant is a miserable place to be. How does someone get there? There are two ways I've seen (nice, capable) people end up in this quadrant:

- They're completely over their head, maybe because they're so lacking in capability (or capacity!) for whatever situation they're in.

- They've been bullied.

Our manager who really wanted to do a good job, well...three months after we last saw them, their boss quit, and they ended up with his job and their own. Three months later, they're managing three distinct teams and 20+ people, and they're constantly being chased from above about everything that isn't going well (nothing is going well). So when an IC complains that their 1:1s are only every other week and often start late, the poor manager doesn't know what to tell them, except that it's not going to change anytime soon.

If you recognize that you're in this quadrant—and I really hope none of you are—then it's time to consider how much of that is situational (e.g., you're in the wrong job; more on that in Chapter 4) and determine how to change the situation.

REFLECTION

When have you been in this quadrant? What prompted it? How did you get out of it?

Applying This to Supporting Others

High actionability with high receptiveness
> This is the quadrant with the least friction, so giving feedback in this quadrant should feel—and be—minor and contain a lot of affirming or validating feedback that what the person is doing is working. Take a curious mindset and encourage self-reflection, and support them in thinking critically about what feedback means and what they want to do with it.

High actionability with low receptiveness
> You have to work to get this person to take feedback, but when you do, they do so much with it that it makes it worthwhile. Work on understanding why this is: are they scarred by too much bad feedback in the past? Do they struggle to trust your opinion (or that you have their best interests at heart)? Work on building that trust with them. Try asking for their opinion, and, in return, show that you value it.

Low actionability with high receptiveness
> This can be frustrating because it may seem like this person has taken the feedback well, but then...nothing happens. Dig into why that is. Are they overwhelmed? Do they not know what to do? Make sure you set time to follow up postfeedback and agree on concrete steps. Then, follow up on them regularly.

Low actionability with low receptiveness
> Coaching people in this quadrant is like banging your head against a brick wall. In this quadrant, you need to be very direct and concrete in what you want to see. Focus on getting a single concrete change, and accept that you may never do more than that.

Becoming More Coachable

In any relationship that involves coaching (with a manager, mentor, or coach), consider how you're likely to hold yourself back and what you can ask for to get the most out of the relationship. If you think what they're asking of you or suggesting is fair, but you don't know what to do with it, ask for concrete suggestions and hold yourself accountable to them. If you're struggling with the content of the suggestions, consider if you can take that struggle elsewhere or be open about why that is.

Like everything else, we can work on becoming more coachable. This is a really powerful catalyst for individual growth with very strong and long-lasting effects. Because coachable people are easier and more rewarding to help, they get more help and do more with it.

So how is that done? Here are five ideas on how to become more coachable:

- Build your self-awareness
- Broaden your perspective
- Shed your defensiveness
- Own up
- Ask for advice

BUILD YOUR SELF-AWARENESS

The most exhausting people to give feedback to are those who are so invested in some image of themselves that you can never really talk to them—only their ego. The easiest people to give feedback to are those with few self-illusions and a level of self-worth that makes it so they don't find it threatening to know what they can improve.

In short, the more self-aware you are, the more people can connect with *you* and not the story you need to tell about yourself. Self-awareness is often hard-won, but professional coaching (if you find the right coach!) can help, as can therapy and good friends who are willing to call you up when you get in your own way.

BROADEN YOUR PERSPECTIVE

Broadening your perspective helps you see things in different ways and be more open to possibilities outside your worldview. There are three good ways to help with this:

- Read a broad array of fiction. Reading (literary) fiction makes people more empathetic.[7]

---

7 Julianne Chiaet, "Novel Finding: Reading Literary Fiction Improves Empathy" (*https://oreil.ly/7UuIN*), *Scientific American* (October 4, 2013).

- Cultivate a broader network of people. Start by expanding the voices you listen to on social media; over time, try and broaden your friendship circle.[8]
- Travel outside your comfort zone (when it's safe and financially possible). Exposure to different cultures, ways of living, languages, and values can be a great way to expand your horizons.

SHED YOUR DEFENSIVENESS

As a rule, I try to never defend myself when someone gives me feedback. Defensiveness either shuts the conversation down or makes it about your feelings rather than what the person is trying to tell you. Try to accept that anyone who cares enough to give you feedback is not setting out to upset you; offer context they might be missing, but remember that too much context is just a nice way to defend yourself. You'll learn a lot more from the conversation if you ask questions and "get curious" instead.[9]

Importantly, you don't have to respond in the moment. You can take time to process—maybe work through your defensiveness with someone else (this is why we need a broader support system, as discussed in "Building Your Support System" on page 33)—and come back to the person to continue the discussion. Removing pressure to respond in the moment can help you avoid being defensive and give you space to decide what part of the feedback is useful to take. Remember, the second score is the most important one.

Also remember to keep your own agenda in mind. Some of your actions may invite people to criticize what you're doing. But what it really means is that the person doesn't understand your long-term plan. If they don't understand the bigger picture, they won't understand the smaller steps needed to achieve that strategy. One of my leads and I had a long-term disagreement about the need for a release train. I felt strongly that we should have one, he felt strongly that we should not, and since he was doing most of the releases, I was doing none of the releases, and there were no obvious problems, his point of view prevailed, and we accepted that we just disagreed on this particular topic. Then one day there was a problem! And we finally had the argument! It turned out that we both thought that if we had a release train, it would get overridden by product wanting

---

8 It's worth noting that you'll probably have to make more effort to become friends with people who are less like you, and you'll often have to do more of the emotional labor.

9 Dani Rukin, a coach I worked with for several years, really retrained me to "get curious" instead of getting annoyed.

to release things on specific days. Which led to him concluding that meant there was no point having a release train, and me still wanting one because it would allow us to surface the cost of these ad hoc releases and make a case for investing in other things—like better feature flagging—that would make that work easier on the team. Getting to the *real* disagreement was really useful, and I wish we had done it sooner.

OWN UP

If you can admit what you're bad at or mistakes that you've made, the conversation starts with what you want to get better at, rather than forcing the feedback provider to convince you this is a thing you need to work on. When you start with a self-assessment that demonstrates self-awareness, a lack of defensiveness, and empathy for how your actions and stress fall on others, people are much more inclined to believe that their feedback will be heard and acted on constructively. Being able to show some of your own areas for improvement can go a long way toward allowing others to feel safe in identifying their areas for improvement.

For example, whenever something doesn't go to plan, I think about what I should apologize for first. Was there something that wasn't clear? Or something that I should have been trying to change? Did I underestimate the impact of something? And then I can start the conversation more like, "I thought the problem was X, but now I realize Y was also a problem, and I should have advocated more for that to change, and I'm sorry I didn't." Then, we can have the conversation about Y: why wasn't that problem clearer? What did I miss? What could the person I'm talking to have been more assertive about? How will we avoid it next time?

ASK FOR ADVICE

People are often afraid to offer unsolicited feedback, particularly upward to their managers or tech leads. They assume that the feedback might upset us, that they have nothing useful to say, or that they lack the necessary context to offer useful thoughts. They're also particularly unwilling to take the risk of offering feedback if they think things are going well overall. Why risk the potential upset? However, knowing the mindset of your team members is incredibly important because how can you improve if you don't know what's going on?

For most people, it's a lot easier to give advice rather than feedback, so ask for that instead.[10] This can be as simple as asking them, "What would you try?"

---

10 A great insight from Claire Lew (*https://oreil.ly/TO1B9*).

Asking for advice works better than asking for feedback because it sets the bar lower and asks people to articulate what they think might improve things, rather than say what didn't go well.

This idea of just asking for advice scales across various levels. When we discussed making the most of 1:1s with skip-level managers back in "Building Your Support System" on page 33, we talked about how a skip-level manager may feel like they lack deep enough context to give actionable feedback, but you can steer the conversation in the 1:1 so that you're asking for specific advice.

One thing we saw during the pandemic, if we come back to drowning rats, was that the water was rising and there were fewer life rafts around. Even when there's nothing we can do about the practical realities of life in "unprecedented times," the purpose of coaching or mentorship isn't to ignore the reality of things as they are but to expand our capacity even as we carry the additional burdens of whatever is going on. We can also apply this thinking to diversity, equity, and inclusion (DEI).

We'll talk more about building diverse teams in Section 3, "Scaling Teams", but often we focus on grand ideas and structural change—don't get me wrong, these need to happen—but miss how relatively small actions on our own part can make a huge difference for under-indexed individuals, and that this can be true even when we ourselves don't yet have significant structural power.[11]

This tweet (Figure 3-3) I sent in 2016 set off a chain of events for my friend Jazbo Beason. I'll let him tell you about it.

cate // @hachyderm.io
@catehstn

For February if you are underrep in tech and work on mobile, @buritica or I will have a call with you about anything you want.

11:11 AM · Feb 4, 2016

💬 4   🔁 39   ♡ 24   🔖   ⬆

Figure 3-3. Setting off a chain of events (tweet from the author dated February 4, 2016)

11 Often we use the word *underrepresented*, but I prefer *under-indexed* as it implies not found rather than not present. Credit to Duretti Hirpa for this.

Story

As fate would have it—faith, my mom would probably argue—this tweet landed on my timeline in 2016.

As a self-taught iOS engineer, I was building and shipping apps, but apart from that, I had no particular predisposition to becoming an engineer. I immigrated from Jamaica in 2005 with the intention of pursuing a career in medicine. When that didn't work out, I bought a MacBook Air, learned to make iPhone apps, and shipped my first app to the App Store. When I saw this tweet, I just needed a shot, a ramp into an entry-level engineering position.

I reached out to Cate, and we scheduled a call. She asked me a few questions and answered mine. At the end, she said she would put me in touch with some folks she knew were hiring. She connected me with Glowforge (*http://glowforge.com*). Cate could have stopped here— most people would have been satisfied with just a referral of this nature after just one call—but where mentorship ends for some people is where sponsorship begins for her. She offered to do a couple of mock technical interviews and asked when would be good for the first one. I said, "Tuesday at 9 a.m.," and she said, "OK, cool" and sent an invite.

That Tuesday, she came online at 9 a.m. Eastern, seemingly very tired. "Are you OK?" I asked timidly. "Yeah, I flew in a few hours ago from Seoul. My alarm just went off," she muttered. "Why didn't you ask for a time that would work better for you?" I asked, a little perplexed. "I know you have a family and job, and I wasn't sure if this was the only time that would work for you, so I made it work."

Sponsorship goes beyond providing guidance and into provision itself. While mentorship is describing what a good launch site looks like, sponsorship is literally fueling your rocket. Mentorship might describe the value of a network, while sponsorship might ask someone prominent in a network to give you a call. There is a cost, of course, but that's the point.

I will never forget Cate's sacrifices, but it is how casual she was about all of it—how casual she still is—that stood out the most to me. "Yeah, sure."

She bought books that helped me level up.[12] She referred me to be a cohost at try! Swift New York in 2016, which paved the way for me to give a talk there in 2017 and then at try! Swift Tokyo the following year. The rest, for my career, is history. After a great stint at Glowforge, I landed a role at Apple in 2020. In 2021, I was promoted to senior engineer. I now work as a research engineer on the Human-AI Interaction team.

It is Cate's conviction that I would have done this without her, and who am I to question her? What I know for sure is that this is my story, my trajectory, and everyone that knows me knows that I did the work, but it is her sponsorship that made the way.

It is funny how fate disregards your position on faith when it does its thing.

—*Jazbo Beason, software engineer*

I gave Jazbo practical help: a life raft. Because that's really what most under-indexed people need when they try to break into this industry. But here's the thing: I did a lot of those calls, and he's the only person I heard from again after the first one. He's the only person who ever came back around and followed up. So each time he came back, I gave him a little more, and each time he multiplied it. I really could not be prouder of him and everything he's done.

I'm human, so I enjoy the credit he gives me, but more than that, I love to see him paying it forward and having an impact on other under-indexed folk. Above all, he's the person who showed me what it looks like to maximize what you get out of every such interaction: how to do the most with the life raft, the swimming lesson, and the view from the bridge. When I think about who I want to spend my time on, I look for people like him who will multiply whatever I give them and really make the most of it, and that's the kind of person I try to be in turn.

There's a lot to unpack here, and the journey of being more coachable is one that we have to revisit in every relationship and experience. If you make one change, make it this one: as you go into any mentoring or coaching situation or your next 1:1 with your manager, consider how you can get the most value out of these interactions. What do you need to ask for? And how can you get out of your own way?

---

12 Specifically, *Cracking the Coding Interview* by Gayle Laakmann McDowell (CareerCup).

Bad Feedback

I truly believe that the work of being more coachable is worthwhile, but it's also important to decide what to take and what to leave. Don't bend yourself into a pretzel trying to be something you're not, and don't invest too much in things that are ultimately coping mechanisms (like minimizing or silencing yourself or perfectionism) that won't serve you—will, in fact, harm you—elsewhere.

Feedback is just information. Sometimes it's information you can use to improve, and sometimes it's information you can use to make informed decisions about where you should be spending 40+ hours a week.

When I had the (extremely miserable) job of scaling hiring, my boss (the CEO) consistently gave me feedback that we weren't hiring fast enough.

And I would try to explain (ideally ahead of getting into/outside these feedback conversations) that we were working on building a *scaled* process that would result in us being able to hire more, and more sustainably, indefinitely. There was a color-coded spreadsheet that modeled everything, that I was making a decision to prioritize the long-term goals over the short-term pressures, and that I was constrained by the level of people management/onboarding in the organization to hire people who could be successful in that.

Eventually, we did build that scaled process. It took around six months, and we started consistently meeting our quarterly goals with some room to grow (and improve efficiency). Some three to four months after that, we got the feedback that the throughput and consistency of the process was the best thing.

Like, duh.

We had so many rounds of the first feedback. *So many.* I tried everything I could think of, including telling him he could have someone else do it (he did not take me up on this).

Ultimately, what I took from that feedback had nothing to do with the gap between the quarterly goals and the initial output—which I was fully aware of. What I took from it was:

- I was not being listened to.

- I was not being supported.

- My "taking one for the team" was not being valued.

- It was time to find another job.

At first, I got ground down by these conversations and went to the place of "I am not enough" and "Nothing I do is good enough." But validation from my external peers—that these things take time, that my metrics were good—and work with my therapist and coach took me to a place where I could see that this feedback I was getting was much more about the person giving it to me than it was about what I was doing. That what I was going to take from the feedback was that *I deserved better*. And once I'd found it, I moved on.

Bad feedback:

- Ignores broader context
- Negates your input
- Undermines rather than helps you succeed

IGNORES BROADER CONTEXT

None of us operates in a vacuum. We work within teams and within organizations, and we exist within structures of oppression.

Sometimes the context is the work: feedback that complains you didn't succeed at X that doesn't acknowledge that you didn't get input Y will not feel helpful. Feedback that helps you identify how you could have realized Y was delayed and figured out options to mitigate or better communicate it would be helpful. In these situations, it can be worth highlighting the "real" issue and asking how you could have better managed that. Get curious and dig into the feedback; once you can identify a change *you* could have made, you've found something to learn from it.

If you can't have that kind of conversation, you'll have to do the work on your own to figure out what to take from that feedback, and it might include finding a better work environment.

Sometimes the context is systems of oppression: feedback that enforces systems of racial or gendered oppression. Put this feedback in the bin.

The risk is this: for example, women are often given feedback on their "tone," but that's really just gender bias. It's fair to put this in the bin. **But there does exist feedback that is about how you present or come across that can be useful to you.**

I have definitely gotten feedback about my speaking that I have a higher-pitched voice and a British accent...what am I supposed to do about that? But I've *also* gotten feedback that I talk very quickly, and I can see how that combined with a higher-pitched voice (less projection) and a different accent (harder for people who aren't native English speakers) would make it harder for people to follow my presentations. There *is* something I can do about that: deliberately work on my pacing. That's useful to me.

It's possible that working on my pacing is really what I was supposed to take from the initial gendered nonsense, and the unfairness of receiving that kind of feedback is we have to figure out how to put it in the bin, because otherwise it will just make us feel bad about who we are and things we cannot change—even if we wanted to.

But it's a shame if we oversort too much into the "bias bin" and lose the opportunities we have to learn and improve.

I'm not suggesting we should mine all biased feedback for gold—don't do that to yourself. Instead, I'm encouraging you to mindfully sidestep the trap where learning how to discard biased feedback means we start discarding too much feedback, period.

And of course, this depends on volume and who it comes from. If you get one piece of feedback from your manager on your presentation style, you probably have time to do something with it. If you have a popular podcast and people complain about your voice constantly, put it all in the bin. You have better things to do.

NEGATES YOUR INPUT

My favorite book on feedback is *Let's Talk: Make Effective Feedback Your Superpower* (Penguin) by Therese Huston (no relation). One of the **key** points in this book is to **listen first** when giving feedback and ensure the person feels heard.

Note

A good feedback giver listens first.
A mediocre feedback giver listens after.
A bad feedback giver does not listen at all.

If someone doesn't listen to you, then the only reason you have to listen to them is their power over you. You get to decide how valid that power is and what role you want it to play in your life going forward.

UNDERMINES RATHER THAN HELPS YOU SUCCEED

When I was reading *No Rules Rules: Netflix and the Culture of Reinvention* by Erin Meyer and Reed Hastings (Penguin), one of the things that initially seemed quite shocking to me was the direct feedback culture. Like, wow! People actually just...say what they think, when they think it, without preamble.

But actually, it turns out that below the headline is the most important point: people are encouraged to only give feedback that will **help the other person succeed.**

This changes everything, right? Even if feedback is hard, if I can see how it will make me better, it will be much easier to take.

If not, well, why is the person giving it? To make themselves seem better?

A pattern I've seen in difficult people (the kind who are a net negative on the team) is that they give three kinds of feedback:

Self-centered or self-serving feedback
"You did what I asked, good job." / "You did something that I agree with, and this is why I agree with it."

Broad, undermining feedback
"This person is just not a good leader." / "This person is not effective."

Petty feedback, either directly or indirectly
"The meeting notes were posted a day late." / "They were 15 minutes late one day." / "They did not reply to this ping in Slack."

None of this feedback is there to make the other person better. At best, it's there to make things happen in a certain way or to enforce certain standards. And yes, obviously it's preferable if people are on time and the meeting notes are posted when they should be. But it's just unkind and wasteful to ding people for every little mistake.

The challenge of this kind of person is that if you take their feedback seriously and look at each piece of feedback individually, it can seem worth listening to. These people often talk a good game about how they're just trying to make everything better (or "preserve the culture"), and maybe they really believe it. But "better" is extremely subjective, and once you figure out that someone's idea of "better" is the one in which they have more power and others have less, put it all in the bin.

Key Takeaways

Sometimes the feedback is much more about the person giving it than you, and what you learn from it is how they think about things and what they value. From their feedback, you may learn how to work with them more effectively or that this is a person to avoid, and that can be a reasonable conclusion.

Sometimes the feedback is hard, but there's something you can learn from it. The work we do on ourselves to untangle previous feedback trauma so that we can keep learning is a huge gift for our own resilience and growth.

Sometimes we're lucky enough to have great, supportive relationships with people who only want to see us succeed. If you have those, treasure them.

4

Moving Forward

In this chapter, we'll look at the signs to pay attention to when it comes to thinking about quitting your current job and how to define an action plan based on where you are now and where you want to go.

Signs It's Time to Move On

In 2020, when the pandemic started, our way of life and work changed dramatically and abruptly. Over 18 months or so of pandemic life, many people had more time to think about what they wanted from their jobs and the kinds of conditions they were willing to accept, the impact of which became known as "the great resignation." For knowledge workers, this conversation has often been about perceived entitlement, such as those who didn't want to return to the office.

Regardless, your current job is just a moment in your overall career, and it's worth thinking critically about whether it's serving your longer-term career goals. Leaving a job isn't necessarily a sign of giving up or of being unsuccessful. Instead, it may be the best thing for your career, and making a move might be a really positive step.

So, here are five reasons why you may want to think about moving on.

YOU'RE NOT LEARNING (AND YOU WANT TO BE)

It's normal to move between periods of higher growth and periods of consolidation, and sometimes it's a relief to be operating well within your comfort zone. But if you're poking your head up and looking around, and your next growth opportunity is nowhere to be seen, it's worth considering that your next growth opportunity could be elsewhere.

The trap

As mentioned in Chapter 1, sometimes five years of experience is just...the same year of experience, five times over. This can really set back your career trajectory, making it harder to interview or get hired for roles you think you *should* be qualified for by now. Employers who interview in depth will suss out if you've been stagnating and will be more likely to pass on you for a "more qualified" candidate whose experience has been more varied.

Before you quit

Talk to your manager about what growth opportunities they see for you. Especially if you've struggled personally for whatever reason, it's worth talking about how whatever was going on is behind you now and laying that period to rest between you, making it clear you're ready to take on more. Many nice and understanding managers may let people drift a bit, not wanting to add pressure when they're going through a tough time; being clear about what you're ready for will make it easier for them to help you get it and know that it's time to start pushing you again.

If you move on

Be intentional about what kind of learning and growth you can really expect in the new role, and be clear about optimizing for it, even if that comes at the expense of other things you care about (like job title).

Story

I had been working in software development since 2007, mostly doing quality assurance, and managed to reach a senior position where I could do both technical work and people management. However, four years ago, I found myself wanting to do something more, still related to software and people but not with testing anymore.

I worried that I lacked the technical experience to move toward engineering management (EM). My husband, who also works in software development and had stepped away from coding toward a more people management role, encouraged me to pursue that path, but in my eyes, I wouldn't be able to make it through a selection process for that position, let alone be able to coach developers on technical topics in which I myself was not experienced.

Regardless, my manager decided to move me into engineering management within the company, so I took the leap from being QA lead/manager to engineering management (hurray!), but since the other EMs in the company were heavily technical, the expectations and the definition for my own role felt different and vague: I was still managing, but only QAs, and I wasn't involved in their projects because they all were assigned to different projects. It was a people management role, but I had no say in the product they were building or the teams in which they worked at all, which in turn would limit the impact of what I could do for my reportees. In the end, I quickly burned out due to frustration and quit without having a new job lined up.

With the help of a mentorship and a lot of interviewing, I managed to land an EM job where I was able to work side by side with product, plus I got to manage a cross-functional team. What drew me to this new position was the opportunity to do more. To run a bigger and cross-functional team and to have input in product decisions. In short: more responsibilities and more trust. I still work there, and although we've had some tough moments, the team, my peers, and my boss still help me learn so much.

—*Margarita Gutierrez, engineering manager*

YOU'RE LEARNING COPING MECHANISMS RATHER THAN SKILLS

Every organization has their quirks that people find their way to work around. Perhaps the reporting is a little overly arduous, or your manager's manager is a little political, or the culture is a little too argumentative for your liking. Over time, we learn to cope with these things: we set aside extra time for the reports, make sure we take the time to sell the political person on our ideas, or learn how to argue.

The trap

Sometimes organizations are (or become) sufficiently dysfunctional that we're investing more time in developing and refining the coping mechanisms than the actual skills. If your list of things to develop is really a list of things that you won't have to do in a more functional environment, none of which will make you more employable elsewhere...it's time to walk away. The coping-mechanisms trap is particularly vicious because the more time you invest in refining them, the more time you're going to have to spend untangling them in a healthy environment—if you ever make it to one.

Before you quit

Talk to someone you trust who won't just support you but will also challenge you. It's important that they have an external or at least dispassionate perspective—someone who's also deep in the same coping mechanisms will be more likely to justify them, or you'll just end up venting together. An external coach, a previous manager, or a close industry friend can all be good people to turn to. Ideally, they can check you on what's bothering you: are you overreacting? Would the grass really be greener elsewhere?

If you move on

Put real thought into what coping mechanisms you have developed and expect to find new ones in your next job. You'll have to commit to untangling them, and that can be hard, so figure out what support you need.

YOU FEEL MORALLY CONFLICTED ABOUT HIRING

I'm not suggesting we should all be a corporate shill, but if you're hesitating mentioning that the company you work for is hiring and you're offering a lukewarm view or even "I don't recommend it" to friends who ask you, it's worth asking yourself: if they deserve better, maybe you do too?

The trap

We tend to consider moving on to the next job much more deliberately than we do considering staying in the one we have. It's easy to keep ticking along because things are "mostly fine," but sometimes the questions that people ask when interviewing can remind us that we don't have great answers to those questions ourselves if we let ourselves think about it. For example, in interviews people will often ask questions around company culture, growth opportunities, and commitment to DEI. It's worth asking ourselves those same questions from time to time.

Before you quit

Is it the company, or is it you? Burnout can make us feel ambivalent about things that we'd normally enjoy. Try taking a real vacation and seeing how you feel.[1]

---

1 In Europe, these last two to three weeks, and emails and instant messages go unseen and unresponded to. This kind of break isn't always possible for everyone, of course, and vacation length will vary by the customs of your country. But even a one-week break can help.

If you move on

> Make sure what you're looking for is realistic. It's easy to think that other companies have a "perfect" culture based on their external presentation, but no organization is perfect in reality. Validate by asking questions that help you understand the realities of people's day-to-day—not just what's on the website.

YOUR JOB IS AFFECTING YOUR CONFIDENCE

The best advice I received early in my career was "If it's affecting your confidence, then it's a problem." It's something I still think about and assess situations against. Something might be annoying and easy to shrug off, but things that erode your confidence should be paid attention to. As a rule, over time you should feel more capable, not less. It's particularly cause for concern when you can look at your achievements and the way you're being treated and see a real mismatch.

The trap

> Once you stop feeling valued and start doubting yourself, it becomes harder and harder to find something else. You're not valued, you don't feel successful where you are, so why would somewhere else value you, and why would you be more successful elsewhere? The truth is that success is a product of personal and environmental factors. Maybe all you need is a different environment to help you thrive.

Before you quit

> Make the time to thoroughly and (as much as possible) dispassionately review the things that have been eroding your confidence: are there things you can learn from and use to genuinely improve? We talked about getting the most out of feedback in Chapter 3, and again, I highly recommend the book *Thanks for the Feedback* by Douglas Stone and Sheila Heen. If it's other things, such as the way your coworkers communicate, consider whether the signs are isolated things you could build some resilience to or trauma from previous bad experiences.

If you move on

> A totally new environment can also be a short-term hit to confidence, as we have to learn how to navigate it and ramp up to being effective. Don't expect a new environment to be an instant solution; give yourself time to adapt.

If you have a good relationship with your manager, you can try asking for what you need. For instance, if you struggle to get the most out of their coaching because you're so worried you're not meeting their expectations that you can't engage with it, try telling them that and see if they can offer some reassurance first.

YOUR JOB IS AFFECTING YOU PHYSICALLY

Stress is physical. At the point where it's noticeable in your heart rate, sleep (or lack thereof), and physical well-being, you have internalized it.

The trap

> The physical effects of stress can sneak up on us, and when you're not feeling well, the stress and overhead of looking for another job or risking your health insurance may be the last thing you want to deal with.

Before you quit

> You know your work environment, and you know yourself—so you know whether it's worth trying to set boundaries and/or build healthy habits. If you draw a line at stopping work at 6 p.m., will it be respected? If you carve out time for healthy habits, will it be enough to make a difference? If it is, I recommend Gretchen Rubin's book *Better Than Before* (Crown) and particularly the Four Tendencies framework as being helpful for thinking about building (and maintaining) habits.[2]

If you move on

> Remember that stress takes time to dissipate and rebuilding healthy habits is a process. Shifting to a new organization can be an opportunity to reset and rebuild healthy habits, but if you're not intentional about it, you may fall back to the ones you had before.

MAKING DECISIONS

Forty hours a week—or, let's be real, more—is a lot of time to be unhappy. Being unhappy at work bleeds into other areas of our lives, affecting our physical and emotional well-being and personal relationships. I'm not advocating job hopping—there are always things you can try to improve your situation, and we'll talk more about those things in Section 2, "Self-Management"—but as a hiring manager, I find that the people who make me saddest are those who have stayed

---

2 This framework is expanded on in the book *The Four Tendencies*, also by Gretchen Rubin (Harmony/Rodale).

in one place too long at the expense of their own growth and overall career. Regularly thinking critically about what you're getting from your environment—and what you're not—is key to sustained and sustainable growth. Even if you have a great manager, you're still the DRI of your career, and abdicating that responsibility doesn't set you up for long-term success. Tanya Reilly, author of *The Staff Engineer's Path* (O'Reilly), has a spreadsheet template (*https://oreil.ly/ CWHc7*) you can use to check in on these factors every one or two months.

And if you're a manager thinking about how to retain people on your team, consider that your best retention play might—ironically—be making it easier for them to find a job somewhere else. Making sure people feel valued and that they're learning and growing in ways that provide value to them personally and to their overall career trajectory makes it more likely they'll choose to stay. It's much harder to trap people when the market is competitive—but that was never a good way to manage, anyway.

Your Action Plan to DRI Your Career

We've covered a *lot* in this section, so here's a suggested path for applying these concepts to your own career. This can be a significant shift in mindset that can take *months*, so make sure you give yourself time and space, and don't get disappointed if things don't change as quickly as you want them to. This list is long, and maybe a little overwhelming, but you're not supposed to go through it all at once—review your next action and give yourself some time to take it, then come back to it again later.

Step 1: Assess if it's time to quit.

1. If it's time to quit, or you think it might be, figure out your constraints versus need.

    a. Do you need to make your current role manageable to give yourself time to look for something else?

2. If you're a "maybe" on quitting, think about what's pushing you to that point and what you could try to improve your situation.

3. If you're a no, great! Less overhead for you. Move to step 2.

Step 2: Look at what options you want to be available to you.

1. Check yourself on whether it's a job or a title. What do you *really* want to be doing in the 40+ hours a week you spend at work?

2. Assess where you are relative to these options.

    a. For those of you who are on track, great—you're done here.

    b. For those of you where that isn't the case, move to step 3.

3. Define your current moment.

    a. What is your current moment?

    b. What do you need right now?

    c. Given those things, what constraints do you have here?

    d. Discuss the constraints with someone you trust—do they have to be true?

Step 3: Clarify the deal you made with your employer.

1. Think about what your employer is renting.

    a. Are you building market value?

    b. Are you undermining market value?

2. Think about what your employer is buying.

    a. Is there anywhere that they're taking more than they're paying for?

    b. Are there any boundaries to redefine?

3. What kind of deal do you want?

    a. What supports your longer-term goals?

    b. What do you need to support your life *right now*?

Step 4: Identify some proximate objectives.

1. Come up with some (one to three) shorter-term goals that support what you identified as your overall career options.

   a. Make sure you really nail the "what" and the "why"—achieving this goal is meaningful to you and will represent meaningful progress.

2. Put together development plans for each of the goals.

3. Discuss and refine the plans with someone you trust.

Step 5: Think about your relationship to feedback.

1. Identify what quadrant you're in for some key relationships.

   a. If it's receptive and actionable, great: what's working, and how did you get to that point?

   b. If it's receptive but not actionable or not receptive but actionable, why is that? What would help?

   c. If it's neither receptive nor actionable, why is this?

   Is there a way to extract yourself from this situation?

2. Think about what you could do *generally* to be more coachable. Pick one or two suggestions from the list to work on.

3. Think about some difficult pieces of feedback you've received. What can you take from them? Is there anything useful here?

   a. Even in the worst contexts, there may be something we can learn about ourselves. And then, it's much easier to let that feedback go.

   b. Some feedback and feedback relationships need to go in the bin. Do you need to put anything in the bin?

Step 6: Audit your network.

1. What relationships are good? Celebrate them.

2. What relationships are OK but could be better?

   a. Think about how things could be better.

   b. Think about things you could do to improve those relationships.

   c. Decide if you want to make some time to invest in those relationships.

   "No" is OK!

   So is "Not right now."

3. What relationships are missing?

   a. Think about what those missing relationships cost you: for example, if you have a very small professional network, maybe that makes it harder to find new opportunities.

   b. Think about what you could do to build more of a network. What would be the easiest things you could do? What would you enjoy?

   c. Assess the gap versus what you might do to solve it, and decide whether or not you want to do it.

   "No" is OK!

   So is "Not right now."

Section 1 Summary

Congratulations! You are now officially the DRI of your career.

Just kidding, you always were—nothing has changed other than now I hope you have some more tools to help you fulfill that role effectively.

In Chapter 1, "Career Decisions and Optimizations", we talked about how to approach career decisions and think about what you're optimizing for. We learned that the work you do is more important than the job title you hold and how to pay attention to the value of your expertise on the open market.

In Chapter 2, "Setting and Executing on Career Goals", we distinguished between career goals and career dreams and learned how to define and execute on career goals that will help you move toward your career dreams or, as we are now calling them, guiding principles. We learned that even if professional development plans are often bad, that doesn't mean the concept is worthless. We also learned how to get the most out of the people around you so that you can achieve the things you want to.

In Chapter 3, "Embracing Growth", we discussed the concept of coachability, when mentors help you and when they don't, and how to become more coachable. A huge part of this concept was about feedback and our relationship with it: how to get it, how to get the most out of it, and how to decide what goes in the bin. We covered what sponsorship is, and why it doesn't always look the way we want it to, and we discussed how to suss out opportunities so that we don't miss out.

In Chapter 4, "Moving Forward", we tied it all together, and we thought about when to move on and how to build a personalized action plan for your career.

In Section 2, "Self-Management", we'll move on to the concept of *self-management*. Many of the concepts we covered here—like the difference between energy and time and evolving our jobs for growth—will come up again but in the context of having a bigger role with more responsibility.

Self-Management

Probably, you have a manager—at least according to the org chart. They may provide hands-on help, guidance, and feedback. Even if that is the case today, over time the scope of your work role will (I hope!) increase, and at some point, your manager won't utilize a hands-on management style; whoever your manager is will no longer have the time or inclination to give that to you. At that point, but ideally before, you will need to figure out how to *self-manage*: how to prioritize your workload, how to improve your own effectiveness, how to get feedback when it's implicit rather than explicit.

Everyone has had a manager they thought was ineffective, and many of us have seen someone (or have been that someone) burn out or "nope" out because they didn't want to do it anymore. It's easy to say "that manager was bad" and much harder to explain why they burned out, to identify what the manager was not doing or shouldn't have been doing. *Burnout* ("a state of emotional, mental, and often physical exhaustion brought on by prolonged or repeated stress" per *Psychology Today* (*https://oreil.ly/V21VD*)) can be something of a catchall explanation.

Management is more about the impact than the tactics, and as a result, many people go into management lacking a clear mental model of what managers *do*. Type A overachievers are inclined to address this by deciding to do *everything* themselves. This strategy can go on for only so long, and inherently, it cannot scale. So what happens when they are forced to admit that approach is not viable?

How do you give to others what you're not getting yourself? You have more options than relying on just one person. So if you want to be a good manager, through some combination of intention and luck, you need to have a good manager yourself. If you want to be a *great* manager, then get great at self-management. The more resources that you have available to you, the more you have available to give your team.

This section is about helping you figure out what it means for *you* to thrive in a leadership role, navigating the mindset shift from individual contributor to leader. How do you manage your energy when that is a bigger constraint than your time? What is your job, and how does it need to evolve to meet the organizational need? How do you identify your leadership style(s), match them to the appropriate situations, and expand your range over time?

5

Energy Management

As an IC, you probably focused on (and were good at) managing your time to deliver on your individual commitments. While time management can still be an issue for managers, often the bigger challenge is *energy management*: managing your energy such that you can be effective in the more difficult parts of your job. We'll also discuss how to avoid the trap of being useful, how to navigate burnout, and how the Maslach Burnout Inventory can help you identify and address the warning signs of burnout.[1]

Time management is an inherently task-based perspective: we look at our calendar (fixed commitments) and figure out how many deliverables (flexible commitments) we can fit in, then order them by priority.[2]

Energy management is an *impact-based* perspective: you look at the impact you want to have and what it will take to accomplish it. You distinguish between your ongoing commitments (the things you have to do consistently) and the one-off commitments, and consider how to balance between them on an ongoing basis.

Time management prioritizes (short-term) output and is a subset of a broader category of "productivity advice"—trying to achieve as much as possible day in and day out. Energy management prioritizes long-term effectiveness and impact.

---

1 This is covered in-depth in the book *The Truth About Burnout: How Organizations Cause Personal Stress and What to Do About It* by Christina Maslach and Michael P. Leiter (Wiley), but a more than adequate summary can be found in John Rampton's *Forbes* article (*https://oreil.ly/Y5JiI*).

2 A topic that is explored in depth in David Allen's book *Getting Things Done: The Art of Stress-Free Productivity* (Penguin).

This chapter is not about productivity advice. Productivity advice is often dispensed as a one-size-fits-all, when the reality is that everyone's productivity system is predicated on how their brain works, and we are all wired differently. We will not be covering time management, although there are many resources out there to help you with it; I highly recommend Laura Vanderkam's work (*https://lauravanderkam.com*).[3] This chapter is about shifting your mindset and setting up sustainable ways of working that will serve you in the long term.

Managing Energy Versus Managing Time

When I coach new managers or transition ICs into management or other leadership roles, one of the initial key struggles (which I remember myself) is feeling overwhelmed. For some, this ends up in exhaustion, and those are the managers who often switch back onto the IC path—they find management unsustainable at that time. Some return to it later; some do not.

Superficially, it's understandable that people become overwhelmed. They get a bunch of new responsibilities and need to process them; it can take some time to filter through it all and figure out how to handle it. The context switching can also be very draining, particularly context switching at different levels of abstraction (e.g., from strategy to detail-level code review).

But something deeper drives this: capable ICs are often differentiated by strong time management. They work their schedule to really prioritize the deep work for some chunk of every day, and they figure out how to churn out small things alongside it. They manage their time well, and that's what makes them more effective than people with comparable technical skills.

However, when you shift into management, the biggest challenge is not time management but rather energy management. This requires a different approach.

As an engineer (that is, as an IC), often the most challenging thing to do in a given week is solving some gnarly problem. Carving out four hours of complete focus to make a dent in it can make a huge difference.

As a manager, often the most challenging thing to do in a given week is holding a hard conversation. It might even be a short, hard conversation. The biggest challenge is psyching yourself up to do it, and while you might (and arguably should) spend time preparing, often that is more about managing your own emotions in order to do it rather than the actual work required.

---

3 Start with *168 Hours: You Have More Time Than You Think* (Penguin). I also loved *Off the Clock: Feel Less Busy While Getting More Done* (Little, Brown).

Even aside from the truly challenging things, in a more "normal" week, being emotionally present in your 1:1s or team meetings allows you to detect potential problems earlier and help people more effectively. When you show up distracted and exhausted, which happens to everyone from time to time (we're human!), you're less effective in some of your highest-leverage activities.

If your meetings are not high leverage, then that is a similar but related problem. If you don't have time or energy to think through how to improve (or delete) them, then that will not change.

Similarly, being proactive instead of reactive is largely an issue of emotional regulation. It requires getting ahead of things, perhaps by doing something very tedious, versus reacting to whatever seems most pressing in the moment. Making active decisions requires a level of emotional calm and mental clarity—mental clarity that we lose when we are stressed and overwhelmed. Proactive work also involves less of a dopamine hit than reacting and "fixing" something in the short term.

If you're a new manager and feeling overwhelmed, the first thing to figure out is whether you have a time management problem or an energy management problem.

REFLECTION

- What things do you add to your to-do list on Monday but delay until Friday? Why do you put them off?

- What things do you struggle to do at the end of the day even though you technically "have time"?

- What do you think is the most valuable thing you do? Why? How much time do you spend on it?

- What do other people think are the most valuable things you do? Why? How much time do you spend on them?

- Audit your calendar/regular work and ask yourself for each thing: how valuable it is, how much time is spent on it, and whether it's energy giving, energy taking, or neutral.

Once you've worked through some of these questions, you will probably know whether the problem is just the sheer volume of things (which is OK because this indicates a time management issue, but you can at least begin to figure out how to remove some of the activities) or whether it is not the volume but rather the emotional drain of certain activities.

The final question is this: when you step away from work, are you able to disconnect? If not, what do you keep thinking about?

As we discussed in Chapter 1, thinking about work long after we have finished for the day—especially in terms of ruminating on things we are stressed about—is how we work 40 hours in time but 60 hours in energy. If we physically leave the computer but don't emotionally leave the work, we exist in a limbo of neither working nor resting, and that creates resentment and exhaustion.

Here are some ideas to improve your energy management:

Think about things you can do at the start and end of the day that ground or center you. This can help you start the day in the right frame of mind.

Even just a 20-minute spin class before I start work makes a huge difference in how I experience the day.

If you don't have a commute, maybe you can replicate the best parts of that in a way that signals the start (and end) of the day—a short walk or a bike ride? A trip to a coffee shop you love?

Before opening the computer (or looking at your chat app), write a paper list of what you *think* is most important to do today.

Think about things you can do during the day to ground you. This can help you stay on track and avoid falling into the trap of reactivity.

I use the Positive Intelligence app (*https://www.positiveintelligence.com*) to do mini mindfulness exercises.[4]

Schedule lunch breaks and take some time to do something away from your desk.

Take a short walk, savor a hot beverage, or read a chapter of a book: what small (shorter than 30 minutes) things work for you?

---

4 The book *Positive Intelligence: Why Only 20% of Teams and Individuals Achieve Their True Potential* by Shirzad Chamine (Greenleaf Book Group Press) is worth reading, but for consistent daily application, the app is very helpful.

Make a daily (or weekly) list that accounts for energy rather than time. This allows you to set more realistic expectations for the week, give yourself a clearer path to accomplishment, and avoid the trap of feeling like you are consistently failing because things "should" be easier than they are.[5]

If you have a heavy roadmapping week or are doing performance reviews, perhaps you won't be in a frame of mind to write a chirpy blog post on engineering best practices. Accepting that and planning for it is much easier than dreading it and struggling.

Look at your calendar and commitments for a day and consider what is emotionally realistic versus technically possible. Try color-coding your meetings depending on whether they give or drain energy.

Take 15 minutes to think about each meeting before you have it, and decide what outcomes you want to drive and how you can do that most effectively. This can help you run meetings more efficiently while ensuring decisions get made.[6]

If you don't have time to do this, you probably have too many meetings. See about deleting some! (Start with the boring ones; they can most easily be replaced by something asynchronous, whether that be email, a chat app, or another form of text.

Consider the activities that you find most emotionally draining. Is there anything you can do to change your relationship with them? This can help you identify where and how you need to better set yourself up for success.

Are you on a learning curve? Do you need to take more time to better understand how to approach it?

Are your expectations for yourself too high? Some things only have to be done adequately.[7]

Do you lack an understanding of the purpose of the activity? Can you find one?[8]

---

5 In coaching, we say, "*Should* is a saboteur word," by which we mean it's driven by your coping mechanisms rather than your best self.

6 Nondecision decisions are to be avoided! But they are a regular outcome of people coming to meetings unprepared and unwilling to commit to decisions and actions.

7 I know this may be a horrifying concept—sit with the discomfort! If you find yourself overthinking something, allowing yourself to be adequate can be hugely freeing and help you get started.

8 It is amazing to me how often I find myself—or someone else—struggling with a very simple task, and the reason is they don't know what it is for.

Is your relationship with the activity shaped by previous bad experience? Can you change it? Catch yourself in old thinking patterns, reset your expectations, and reinforce to yourself that *this* relationship is not the same as a previous bad relationship.

Do you need to be doing this activity? Can you transition it to someone else or just stop doing it?[9]

If we want things to be different, we have to create space for them to change. Being overwhelmed each week and finishing behind and exhausted is a vicious cycle. While things may improve "naturally" over time as you get more comfortable with the role, much stress and suffering can be alleviated with some metathinking into how you're approaching your work and adjusting self-management to emphasize energy management as even more important than time management.

The Trap of Being Useful

You may have noticed that in a nice restaurant, everyone has their role. There is someone handling people as they arrive, other people taking orders, and yet more people bussing plates. You may also notice someone standing around making sure everything keeps moving—the person who runs "front of house." Every so often they do something, but they do not run around; their job is to keep the system running. (Don't worry; we'll tie all of this together.)

Now imagine someone in this role: they are supposed to run front of house, but they are uncomfortable with all the standing around. They want to be *useful*. They have done everyone else's job before, so they know what to do and are poised and ready to pitch in. When the busser has too many plates to carry, they jump in. When two tables want to order at once, they're there. When the greeter calls in sick, they figure they can take over greeting people too.

Maybe filling in like this works for a bit on quiet evenings, but come the weekend, it means nobody is running the front of house and managing things; instead, they're running around doing a bit of everyone else's job...badly. No

---

9 In a remote context, in-person "meetups" can be a big deal. This was very stressful to me because when I joined a remote company, even though I traveled nearly constantly, I didn't organize my own travel; I had someone who handled those details for me. Anyway, I understood that it was my responsibility to organize a meetup, and I tried to do it...which meant that I booked a hotel for the wrong month on my personal credit card. It was—eventually, thankfully—resolved, but I learned an important lesson about accepting my weaknesses and have consistently refused any part of meetup organizing ever since.

one really knows what to expect from them because they are here and there and everywhere, and obviously so busy that it would be unfair to expect anything specific. But also, is anyone going to do anything about the fact that the kitchen is slow? And that people with reservations had to wait because previous guests at the table they were supposed to have are still waiting for dessert? And aren't there supposed to be two bussers but somehow one of them is rarely to be seen?

If you visited this restaurant for a nice meal, you wouldn't care how hard this person was *trying* to help. You probably wouldn't even notice. You'd be too busy writing your review about how long you waited for a table, that the food was cold, and that you won't be coming back.

A great front-of-house person choreographs the restaurant: they see problems before they happen and keep everything running smoothly. In our example, this person had the power to make the restaurant great, but they gave all that power away by running around, focusing on immediate details and never the big picture. Now the restaurant only has three stars on Yelp.

What does this have to do with management? Well, new managers often present as exhausted and overwhelmed, and the question I generally ask in this situation is "How much are you trying to be useful?" Instead of managing, they're too busy focused on trying to step in and attempting to help (badly), which leaves a gaping hole in the infrastructure.

More Like Dave

In my current team, we use a shorthand phrase: "More like Dave." Who is Dave, you ask? Dave is my hypothetical replacement should I run away to join the circus.[10] Dave makes more money than me, does half as much for his directs, and expects them to do twice as much for him.

We all kinda hate Dave. But we also recognize that Dave has some ideas that we would benefit from. He pushes work down, expects more from his directs, and doesn't get caught up helping them; he protects his time to be strategic (or just to take a lunch break or log off at the end of the day). So we look for our opportunities to be "more like Dave," and it pushes us to delegate a little more and do a little less.

---

10 I heard this from BJ Wishinsky as a nicer and less violent alternative to "hit by a bus."

The desire to be useful comes from a good place, often tied to some idea of "servant leadership." People internalize this idea that they exist to work for their team, and they do this by picking up all the small or annoying tasks or grunt work, running all the meetings, planning all the team activities, tidying up the bug list, and so on.

This approach results in the following outcomes for the manager:

Wholly reactive
Unable to focus on bigger or more impactful work.

Buried in small details
Unable to step back and see the bigger picture.

Exhausted
Running around all day picking up after people does that to you.

Overwhelmed
See also: reactive. By being buried in the details, you don't have time to identify and drive structural improvements.

Worse, these managers often start thinking it's their job to make their team happy. Wrong! It's your job to make your team *effective*. Constant picking up of small things does not make your team more effective; noticing the patterns and improving the processes or the projects themselves does that.

This is not to say that we have no responsibility for people's well-being; we do, but it's worth recognizing that we can only create the conditions such that people *can* be happy and effective. Taking responsibility for people's emotions is unrealistic—and disempowering.

The biggest problem I have with the idea of servant leadership is *you can't be the servant of people you have power over*. So either you deny that power dynamic or you give away that power.

Because if you behave in such a way that your team starts to believe your job is to make them happy, what is going to happen when you inevitably have to disappoint them? When you have to tell them they didn't get the promotion, or the project has been canceled, or there is no budget for the thing they want?

They will blame you.

Everyone on your team is—hopefully—an adult. You don't need to "protect them" from the realities of the workplace. Trying to do so is patronizing and a recipe for making yourself miserable. If you believe people are adults, you can help them have clarity and focus: make it clear what's important and what's valued and what's not (more on this in Section 4, "Self-Improving Teams"), and trust them to take their share of team housekeeping.

I'm not saying you should refuse to do anything—I'm saying you should not do *everything*. I'm also not saying you shouldn't help your team—that is an important part of your job. It is your job to help them through kindness but not to erode everyone's effectiveness with niceness (we'll talk more about this in Section 3, "Scaling Teams").

Again, wanting to be useful often comes from a good place: the desire to help. This is something to honor. However, it can also be a product of some level of insecurity if you think that by doing small tasks, you are providing value. It's a way to get the small dopamine hits that we miss from writing code. This is something to critically evaluate.

REFLECTION

Take action to escape the useful trap:

- Pay attention to how often you are trying to prove you're useful. Make a list of things you do when you catch yourself being useful.

- After a week or two, evaluate the list. How much is on it? What has that cost you in terms of focus time or energy?

- Redefine your priorities. What are your most important things to do?

- Catch yourself before you do the thing. Let that thing go undone (by you). Does someone else pick it up? Does it need to be done at all? What would Dave do?

Think about some other ways you can get some dopamine:

- Celebrate the wins that *are* your job: when the team ships something! Or you see someone do something well that they previously struggled with.

- Keep a list of the problems you preempted.

- Pay attention to the impact of your 1:1s (we'll discuss more about making 1:1s useful in Section 3, "Scaling Teams").
- Write a script to automate something tedious.[11]

Remember that even if you can successfully run around in the details of a team of five, there is no way to succeed in this mindset with five teams of five. This can be one of the biggest ways that people get in the way of their own advancement, so the time to readjust your approach is now.

Navigating Burnout

The Maslach Burnout Inventory (previously mentioned at the beginning of this chapter) is hands down the most useful tool I've found for understanding signs of burnout in myself and managing people through burnout. It's an index of six causes of burnout—only one of which is overwork—which helps frame the conversation constructively and makes it less personal.

The six causes of burnout are:

- Lack of control
- Lack of reward
- Absence of fairness
- Lack of community
- Conflict in values
- Work overload

There are two key problems with the way burnout is typically discussed. The first is the focus on overwork as a cause, leading people to miss other, more pernicious causes. The second is the focus on burnout as purely a work problem; these causes are generic, and we can experience them in multiple parts of life other than just at work.

---

11 When I was first moving away from coding day to day, my spreadsheets got progressively more intense; you can basically code in a spreadsheet, but no one else will get so excited about that spreadsheet that they add it to the critical path.

There are three steps to navigating through burnout:

1. Understand the context: whether it's a work problem, a life problem, or some combination of the two.

2. Identify the causes: use the index to understand the key drivers.

3. Work toward change.

UNDERSTAND THE CONTEXT

It's important to frame burnout conversations as going beyond the workplace. For starters, often what makes people realize they are burned out is the impact on their personal lives. But second, when people are struggling in their personal lives, they often turn to work as a place where they feel like they have more control, and then they project their frustrations onto their job.

Not unrelated, the best predictor I've found of someone who can survive in a highly dysfunctional environment is the quality of their personal life. Deep relationships and engaging hobbies help people disconnect because they are *connecting* to something else entirely, rather than just trying to separate themselves from work. I have come to believe that we don't find work-life balance by doing less work; we find work-life balance by doing *more life*. Ultimately, these things both come down to the core of work being just part of our lives and the need for some semblance of balance for things to feel sustainable.

One suggestion for a practical first step is to remove work as a topic of small talk: stop asking people what they do when you meet them.[12] Much as I love having tech friends who *totally get it*, I also love having nontech friends who have no concept of what it is that I do. It breaks any temptation to complain about small things; we talk about the other things we have in common instead.

Higher level, allow yourself to consider the bigger picture where your work is just one part of your life. Often, we don't allow ourselves to feel how we feel because our problems seem small in the context of what's happening in the wider world, and we think we are "supposed" to be happy, or grateful, or whatever. What's really going on? And what do you really need to support how you navigate it?

---

12 Credit to Jill Wetzler for this suggestion.

IDENTIFY THE CAUSES

Work through the list of causes and take time to consider each one:

- Lack of control
- Lack of reward
- Absence of fairness
- Lack of community
- Conflict in values
- Work overload

How much does each cause resonate? Where—in life and work—does it apply and to what extent?

This process can be such an opportunity to learn. It may turn out that lack of community is a huge factor, and that is really more of an outside-of-work issue. It can be the case that lack of control in your personal life is making any lack of control at work even more of a struggle. Growth and feeling valued may have morphed into resentment about not getting promoted, or some particular process is having a disproportionate impact that wasn't obvious until there was deeper thought. Once you've identified the biggest causes, you can decide what to do next.

Try this as a writing exercise or with a coach or close friend. As you work through each cause, take the time to really consider where it shows up, both at work and in the rest of your life, even if it's not a big deal for you or doesn't happen regularly.

A Note on DEI

It's important that we talk about the relationship between burnout and under-indexed people. The correlations between the causes of burnout and the experiences of people who are in the minority are glaring:

Lack of control
> Even if you find a "good" environment, that can change very quickly—look at how Twitter's commitment to DEI evaporated when it was sold.

Lack of reward

The pay gap is the most obvious example of this; per the 2022 US Census data, the average Black woman makes 58 cents on the dollar compared to the average White man.[13]

Absence of fairness

There are abundant examples of this. Men tend to get promoted based on their potential, while women get promoted based on past experience.[14] Black workers get more harshly judged for mistakes.[15] Layoffs disproportionately affect women and people of color.[16]

Lack of community

It's much harder to have a sense of community when you are one of few.

Conflict in values

This often shows up when public commitments to DEI are undermined by actions.

Work overload

Women are far more likely to be burdened with office housework.[17]

Sometimes, a sense of burnout comes from broader industry trends or specific bad experiences. It is also possible that some unaddressed (or poorly addressed) instance of racism or misogyny can cause an ongoing dissonance around conflict in values. Often, we talk about the extent of

---

13 This statistic and, more important, suggestions for how both Black women and allies can address it come from "How Black Women Can Navigate Pay-Gap Gaslighting" (*https://oreil.ly/cz3iz*) by Lola Bakare, *Harvard Business Review* (September 21, 2022).

14 See Katie Bishop, "Proof Versus Potential: Why Women Must Work Harder to Move Up" (*https://oreil.ly/slOKn*), BBC, February 22, 2022. There's less data for people of color on this topic, but the workings of structural racism would suggest similar phenomena apply.

15 Gillian B. White, "Black Workers Really Do Need to Be Twice as Good" (*https://oreil.ly/NGkzd*), *The Atlantic*, October 7, 2015.

16 Layoff data combined with name analysis indicated higher percentages relative to employment percentages; for more, see Jared Lindzon, "How Recent Tech Layoffs Can Disproportionately Affect Women and People of Color" (*https://oreil.ly/HMOMy*), Fast Company, January 25, 2023.

17 Office housework includes things like planning meetings and arranging social activities, work that women are punished for if they refuse but get no credit for doing. For more, see Deborah M. Kolb and Jessica L. Porter, "'Office Housework' Gets in Women's Way" (*https://oreil.ly/T5Yyi*), *Harvard Business Review* (April 16, 2015).

what happened, but the message sent by the broader organization may have had the more negative impact, and if it affects you, it affects you.

If you're an under-indexed manager, honoring your feelings and the impact situations have on you is important. If you manage someone who is under-indexed, consider how these things may have affected their ongoing experience.

Often when we talk about DEI, we talk about the "second shift." This idea is that the first shift is the actual job (what we are hired and paid for), and the second shift is the DEI work under-indexed people get pushed into. I believe we need to talk about the third shift—of self-healing. Inequity is by definition inequitable—unfair. The first unfairness is what happens, and the second unfairness is that the only person who can understand and try to untangle the impact it has on you is yourself.

WORK TOWARD CHANGE

Hopefully, by now it's clear that unless the cause is pure overwork, *unstructured time off is unlikely to help*. In software, hiring is expensive, and many companies will make some effort to retain people. One of the easiest levers for this is giving them time off and hoping they will come back. Sometimes this works; often it doesn't.

Not that time off is never helpful—it often is. However, to change things, you need to know what you're trying to change and why, and structure time off in support of that. A month of four-day weeks may be more useful than a week off. Several one-week breaks with time to plan them can be more helpful than a month off starting tomorrow. A concrete plan to switch teams with time in between for a break can be more useful than going away and coming back to a situation that hasn't changed and doesn't seem likely to.

Often, the biggest problem with burnout is that people don't know what they need and why. They feel ready to flip tables but haven't extracted the driving reasons for that, and if they feel bad enough, a huge change like a new job or six months at the beach feels easier than trying to change things where they are.

However, once you've identified the causes, you have a structure for creating change. You've identified the drivers, so now you can start to identify what *can* change. It's continually amazing to me how much can change if there is space to change it. Once the problem is "lack of reward," it's much more tractable than "I'm annoyed about everything."

In the case of **lack of control**, you may want to focus on what can be controlled and other ways to increase feeling of agency.

In the case of **lack of reward**, you may want to focus on either aligning activities with what the organization values or better surfacing invisible work and tying it to business outcomes.

In the case of **absence of fairness**, you need to either dig in and understand it (so you can self-advocate) or make some bigger changes that illustrate your point. If you think it's unfair that someone was promoted and you weren't, get curious about what that person is actually doing—and what you can learn from it. For instance, are they better at talking about their work? Instead of resenting it, learn from it. Are they tackling hard problems? Look for hard problems you can tackle yourself.

Something I've seen repeatedly in this category is (newer) managers presenting as burned out and no longer wanting to be managers, especially after dealing with challenging performance management problems without adequate support from their manager or HR. To them, it feels profoundly unfair that they feel miserable after someone else hasn't done their job properly. Opting out of the management track is one way to tackle the absence of fairness, and it's a way to make a point about your own experience, but it is disempowering to make a decision based on what you're avoiding rather than what you're going toward. Think honestly about what you can learn from this experience, and—ideally, before it's too late—find someone who will help and support *you*. We'll talk more about performance management specifically in Section 3, "Scaling Teams".

In the case of **lack of community**, you will need to understand where and how to build a greater sense of community—for instance, by connecting more with peers who are internal or external to the organization. It can be *hard* to find community as a senior leader; at this stage, a couple of close friends in similar positions will be more helpful than more generic employee resources or community groups. If you already have a community, great! You know who to spend more time with. If not, don't despair! Look around your network and try to build more of a connection to some people you already know somewhat; it's likely that even a few of them would also like to build out their community, too.

In the case of **conflict in values**, you will need to figure out what is intractable and what is not. As a leader, you can often create a team that mirrors your values. For instance, you can create a team that values teamwork and support in an organization that skews toward the individualistic. However, when you go outside of the team, you will be met with the broader values of the organization, which may not match your own or what you have instilled in your team. Organizational values are not intractable; they can and do change over time. For instance, as organizations grow, they have to become less individualistic and more interdependent. The key is knowing what can move on what time frame, what probably won't, and how you can manage the impact that has on you.

When trying to figure out what can change and what will not, think about the current pressures of the organization, or ask your manager what they worry about and what changes they think will come. Look at the company goals and think about what will need to change to support them, and consider whether those changes are happening and how you can support them. Pay attention to how things evolve. When I got frustrated about things, my work BFF Eli Budelli would often tell me, "That will fall under its own weight," and now I think about that a lot. Some changes really are inevitable—but inevitable does not mean quick, so I had to learn to be more patient.

Work overload invariably comes down to the need to do less and identifying (1) what to drop, (2) what to delegate, and (3) what to do *less well*. Number 3 is critical because often people create their own work overload by holding themselves to unreasonable standards (as we discussed in "Managing Energy Versus Managing Time" on page 84). At times, especially during a period when the organization is growing or there is a seasonal busyness, some things cannot be escaped but should be done no more than adequately.

Managing Burnout in Others

Burnout doesn't show up only in ourselves; we also find it in the people we manage. Managing people who are burned out can be incredibly draining: they think everything is terrible, it's difficult to get them to be constructive, and it's hard to (want to) help someone who is blaming you. The framework can help with this, as it makes the conversation less personal and can help you make it constructive:

Understand the context

Boundaries are important, and I'm not suggesting you start meddling in people's personal lives or assuming their personal lives are the cause of problems. In a reasonably functional, reasonably humane workplace, though, it's worth asking the question of what's going on generally before moving forward. During the pandemic, I would regularly see people exhibiting symptoms of burnout and negating them because "work is fine." Drawing attention to that and allowing them to open up—if they wanted to—let people create change and let me help them create space for that change, whether it was adjusting their workload, setting up a (temporary) four-day week, or taking a longer vacation. Where that undercurrent of dissatisfaction continued to fester, inevitably it eventually affected work.

It's invariably been a relief to me when someone opens up about something big going on outside of work. Even if they try to mentally compartmentalize, there's inevitably been some impact that is explained once I know what they're going through.

This is why this context is the framing for the next part: the causes. By putting wider life and impact on the table, you can more holistically evaluate causes and impact.

Identify the causes

The structure of the Maslach Burnout Inventory is such a gift here. It allows you to take people through the causes one by one and talk about which ones are most present for them. You can probably take a guess from previous conversations which ones are likely to be most present, but I would encourage you to structure the conversation and get curious about how the different factors resonate with the person you're talking to.

Work toward change

As a manager, except in the case of work overload, you can only give options; the person will have to make the changes themselves.

In the case of **lack of community**, you likely need to focus on team culture and engagement. The person may need to be assigned a more collaborative project, to mentor (or be mentored) more, or to be supported to participate in organizational or external communities more.

Conflict in values is interesting. Sometimes this is intractable: the person has values that are orthogonal to those of the organization and is unlikely to be happy there, such as someone who deeply values status in a flat organization (or vice versa). Sometimes someone has the potential to learn a lot and become more effective from their value being challenged, such as the person who deeply values code quality in an organization that values shipping fast. Sometimes people perceive that something is valued more just because it's more visible; for instance, they may view diving saves as more valued than incident-free deployment. In this instance you need to (1) work out whether the value is intractable, (2) understand how (or if) it's serving the person and how willing they are to be flexible, and (3) create a more nuanced understanding of the organization's values where appropriate.

One useful way to discover people's values is an exercise called "value mining," which is essentially asking questions that elicit information about what is really important to someone. *Dare to Lead* by Brené Brown (Random House) has a helpful values sheet (*https://oreil.ly/7dnJn*) and asks you to pick just two(!).

The most obvious place to find someone's values is in conflict because people tend to see their values as a moral good or bad. For instance, someone who really values trust and delivery might get upset about a nitpicky code review that doesn't allow them to merge. But it might turn out that the code reviewer really values being thorough and leaving things better than they found them. Understanding the deeper motivation can uncover the real conflict and make it easier to address without forcing someone to compromise something important to them. The next time you find that someone is annoyed about something, try extracting a value that explains the annoyance.

GIVER BURNOUT

A final flavor of burnout comes from Adam Grant's book *Give and Take: Why Helping Others Drives Our Success* (Penguin). There are three types of people: givers, takers, and matchers. Givers...give. Takers...take. And matchers enforce fairness: they give to those who give to them.

Overrepresented in both the highest and lowest performers are...the givers. When givers are at their best, their willingness to help creates a rising tide that lifts all boats. However, when they are not being effective or are being taken advantage of, they are more likely to become ineffective and even burn out.

The person who is complaining that they are doing everything for everyone else and not being appreciated but who never asked for help? That's a classic giver burnout situation.

Give and Take is well worth the read for this specific situation, and it also references the Maslach Burnout Inventory, but the short summary of the approach is to (1) prevent situations where takers take advantage of givers; (2) look for the giving situations that are win-win, where they benefit both the giver and the receiver; and (3) look for opportunities to scale giving, such as shifting from many 1:1 pairing sessions to group presentations or improved documentation.

Better Energy Management

Hopefully, as you made your way through this chapter, you found places where you can improve your energy management and see how it can be transformative for your well-being. The amount of change you want to make here may feel particularly overwhelming if you're exhausted, so if that's the case, try choosing one small change to make next week. If it works, you'll be encouraged to add more changes. If it doesn't, don't give up, but consider that maybe it wasn't the right change to make *right now*, and you need to choose something easier and more tractable instead.

If you're fine overall, it's a great time to experiment with things that feel bigger and riskier! The more you can improve your energy management, the more you will be able to take on new challenges, as we will discuss next in Chapter 6.

6

Defining and Adapting Your Role

Congratulations, you're a manager. But wait! What even is your job now? What are you responsible for? What do you do? You've read *The Manager's Path* by Camille Fournier (O'Reilly)[1] and scheduled a bunch of 1:1s, and maybe you have some ideas about what needs to be different for the team to do better.

It's easy to get caught up in the activities of management: the meetings, the escalations, the bureaucracies. But it's not always clear what these are meant to accomplish. This can lead to believing you're a good manager because you do all the activities but being unclear on how to progress because you're focused on tasks rather than outcomes.

This chapter is not about the tasks associated with being a manager. This is about the outcomes managers need to drive to be effective. You can use this information to manage down, making sure you drive the outcomes your team needs, and you can use this information to manage up, ensuring you get what you need to be successful.

This is an opportunity to think about what you should be doing now so that you're doing something else six months from now. What does success look like, and how do you know when you have achieved it? We'll come back to this in Chapter 14.

In this chapter, we'll cover the components of managing a team, how to decide what to prioritize, and how to ensure that your job changes regularly.

---

1 Excellent book!

The Components of Management

Often when we talk about managing people, it's synonymous with managing teams. We manage people in the context of the teams they are in: we set direction, give feedback, and build alignment. Managing individuals is often about aligning them with the team they are on and making their role clear and correct in that context so that they are meeting the agreed-upon standards.

There are four main things that everyone needs (in varying amounts), and by default, we expect those things to come from the manager, although some of them may come from people in other roles, like the tech lead or product manager:

- Direction

- Feedback (loops, implicit)

- Practical help

- Support

We'll discuss these four main components first, then turn to managing up and managing down.

DIRECTION

This is where people are heading, what's important, and why it's important. The bigger picture around their work, and how their work fits into it. Part of this is giving people necessary context such that they can set direction (or at least the next steps) themselves.

For less experienced people, it is often the case that you set direction with minimal tolerance for deviation, but more senior folk expect to set their own direction having taken input from multiple sources. For example, if a junior engineer has abandoned their project for a side quest, you probably need to dramatically reorient them. If a director has taken on an additional thing you don't fully see the value of, you can register your opinion but ultimately let them decide for themselves.

Note that you are not always the person best suited to set direction. In my current role, most of the people reporting to me are director level and spread across the organization, including in areas I don't have other involvement in. As a result, the CTO does some level of direction setting for some of them—and then I help them add fidelity to that. Metaphorically, the CTO gives the compass setting, and I provide the topological map.[2]

FEEDBACK

We talked a lot about feedback in Chapter 3. But here, I mean *feedback* as a neutral concept: your work reflected back to you. It includes what is going particularly well and what can be improved, and it illuminates places where you lack context that you may have missed.

Explicit Versus Implicit Feedback

When I'm interviewing managers, I love to ask them to give me an example of some implicit feedback they received and how they actioned it.

What is implicit feedback? *Explicit* feedback is what people tell you directly. *Implicit* feedback is what you observe from what people *don't* tell you or the things they complain about:

- If people tell you they feel a lack of direction, the implicit feedback is that the strategy and roadmap are not clear (more on this in Section 4, "Self-Improving Teams").

- If you have a string of bad releases, there's some feedback there about your development and/or release processes.

- If people aren't clear on something you think you've explained, that is implicit feedback on how you explained it.

As we move up the org chart, we get less and less feedback, and the feedback we *do* get is often less actionable. This is because fewer people have insight into our work: at certain levels (like director or VP), you exist as an interface to abstract some part of the org from *your* boss, and while they may tell you things they notice or anything that makes its way up to them, they are rarely going to get in the details you should be handling. Then, the people below you may feel uncomfortable handing feedback upward, and even those who do will have limited context into the rest of what you do. One product of this is that feedback often misses what we were trying to achieve—for example, when a project someone in your org

---

2 This is a reference to Chapter 2 in *The Staff Engineer's Path* by Tanya Reilly (O'Reilly), which I highly recommend.

is running is totally off track and needs to be dramatically reoriented, and someone has "feedback" (aka a complaint!) about you swooping in from above to reorient the project. It's possible you could have handled that better, but really the problem is that you needed to do it at all.

Implicit feedback is the antidote to this. For a leader, the best source of feedback is mining for it from what happens—and doesn't happen—around you. It's how you find the problems—like the projects that are off track—earlier, when you can intervene less dramatically.

When searching for implicit feedback, ask yourself these questions:

- What did I expect to see but didn't?
- What did I think would improve but hasn't?
- What questions did I get that surprised me?

PRACTICAL HELP

Depending on where someone is at, they may need varying amounts of practical help. New hires or people taking on new and bigger responsibilities will need more as they navigate that change. People who are thriving will likely need very little. We all need practical help sometimes, though, so even with the most effective people, it's worth keeping it on the table.

For example, one of my—highly competent—directors was unable to fill in an HR form; he was wildly overthinking it because it made no sense to him. It took me eight minutes to do it for him. Was the form a good use of time for either of us? No. But eight minutes was a reasonable use of my time to extract him from a situation that was generating cognitive overhead. It's easy to assume you have nothing to offer someone competent, but competent does not mean infallible or superhuman; we could all use some help sometimes.

SUPPORT

This is for people's emotional needs. If they're doing something hard or going through a difficult time, or they are new and need to build their confidence, they may need more support. Try to make the things people are overthinking more tractable, or validate that the things they are finding hard *are* hard (but they can do it anyway).

Some people need relatively little support. In the context of managing up, especially when reporting to founders, we might not get any of it and have to figure out how to get those things elsewhere.

MANAGING DOWN

Make a list of your reports and for each one, ask yourself how much of each of these components you're providing. Are you getting any implicit (or explicit!) feedback that any of these things are lacking? Are you noticing places where you could help but they don't ask for it? Do you need to make your willingness to help more clear?

At the other end of the spectrum, are there any places where you are giving too much of any of these things? What is driving that?

Are you giving too much practical help because it makes you feel useful, but actually, it's making them dependent on you? Are you providing support they should really be getting elsewhere in their life, like from their friends or a therapist? Are there boundaries you need to set?

Are you continually reiterating direction they don't want to accept? Are you giving feedback repeatedly that they are not acting on? Do these things need to be escalated?

It's worth asking: what do you owe to someone? If someone is resistant to some or all of these things, how hard should you push to give it to them? Here are some heuristics that can be useful:

- With new people, their expectations will be set by their previous environment(s), and you will need to align their expectations to the new environment. When I have new directs, they consistently expect far less from me than I think they should, and it takes time to get them to expect more.[3]

- When people have thrived but are now struggling, it's worth spending genuine time and effort to reorient and reset, potentially by using the conversations about burnout in Chapter 5.

- If someone cannot fulfill the responsibilities of their role, you do not serve them by Band-Aiding and resenting it. Having clearer conversations and moving toward resolution are far better for you and the team. We'll talk more about performance management in Section 3, "Scaling Teams".

---

3 My current manager gave me this exact feedback; it took me a *long* time to internalize.

- Everyone should need and get some amount of all of the components, but be conscious of what the normal and expected range of support should be and when people are consistently asking for or demanding more than they should. Difficult people often ask for too much from their manager, teammates, and anyone else they interact with. Even minor feedback loops become a production and a time suck. The payoff in these situations is rarely worth the effort. You cannot rehabilitate people who do not want to change, and the sooner these people are gone from the team, the better for everyone.

MANAGING UP

Work through each of these components, and consider what you get from your manager and how you can get it elsewhere. Note that while there are a lot of suggested questions in this section, I don't recommend asking them all at once as it may make your manager feel like you are interrogating them. Focus on one area, and ask one or two questions in a 1:1. See what you learn and figure out what structures you can put in place to keep those things on track.

Direction

This is the hardest thing to replace. If your manager is not inclined to give you direction, consider asking questions to elicit it—they probably know things they haven't told you:

- What is the most pressing thing for the team to accomplish this month/ quarter/year?
- How does {the thing we are working on} fit into the broader strategy of the organization?
- What core business metrics do you pay most attention to?
- What is your sales pitch to new hires about the value of this team?
- What do you worry about?
- What teams are we most dependent on?
- What does your manager worry about?

If that doesn't work as much as you would like, look for other places where you can get further insight:

- Bring these questions to your next skip-level 1:1.
- Attend all-hands meetings and ask questions if you can.
- Product people need to understand these things to be effective, so they can be great people to turn to.
- Recruiters also need to know many of these things to sell roles, so they can be a useful source of information.
- Employee resource groups (ERGs) can be a way to build connections with people whom you might otherwise not have access to.

Feedback

If you can't get actionable feedback, focus on surfacing implicit feedback:

- Do you have any advice for me?
- What makes you worry about a project?
- What kind of impact do you think this team should have?
- What kind of impact do you think I personally should have?
- Is there anything you would like me to pay attention to?
- Is there anything I can help with?

Practical help

If you find you need to manage up extensively, focus on things that you literally cannot do yourself and make it very easy and specific: for example, if you want them to unblock you on a dependency, resolve something with another team, reiterate something with one of your directs, or make a case for head count or promotion.

Support

This is the most easily replaced component; it's more productive to focus on building a support network elsewhere (more on this in Chapter 2).

The Role of Strategy in Management

Many writers on strategy seem to suggest that the more dynamic the situation, the further ahead a leader must look. This is illogical. The more dynamic the situation, the poorer your foresight will be. Therefore, the more uncertain and dynamic the situation, the more proximate a strategic objective must be. The proximate objective is guided by forecasts of the future, but the more uncertain the future, the more its essential logic is that of "taking a strong position and creating options," not of looking far ahead.

—RICHARD P. RUMELT, *GOOD STRATEGY/BAD STRATEGY*

I love the book *Good Strategy/Bad Strategy* (Profile Books), the core point of which is that good strategy is "simple and obvious." The unfortunate side effect of this is that, while management is often a strategic role, we miss the strategy that looks "simple and obvious." The leader who announces some grandiose vision, hires aggressively, and ships slowly with some dramatic incident is more easily seen as "strategic" than the leader who quietly understands the needs of the organization and sets up the structures required such that the team can meet their part in them, whose releases are largely without incident. As such, organizational scaling issues are seen as "growing pains" rather than strategic defects.

Growing pains are—at least somewhat—real. Any organization going through growth will make mistakes and run into bottlenecks. However, those things will be so much more prevalent and so much worse in organizations that don't anticipate and plan for them.

Applying the playbook that worked elsewhere is not a strategy: a common failure mode in leaders who are hired in.

A good strategy is a hypothesis of what will work based on functional knowledge and your knowledge of your own business—this is a crucial insight. Many people find success in one area, and then fail in the next because they apply the same strategy in a different context. Good strategy is only good in context.

—RICHARD P. RUMELT, *GOOD STRATEGY/BAD STRATEGY*

When developing the strategy for your team, you will need to do the following:

- Understand the current realities of the team and business
- Identify the time frame about which you can reason
- Identify current bottlenecks and prioritize the most pressing
- Identify future bottlenecks and start planning for them

Developing a strategy is work that does not happen in 1:1s or, often, in meetings at all. It is the work that happens when you have time to sit and think about things, when there is nothing on fire or calling your attention—or when you ignore the fires and the calls for your attention because you know that you can't get out of being reactive by reacting but only by developing a strategy that acknowledges the realities of where you are at by constructing a realistic plan for something better.

It's the work that is easiest to drop, especially in pursuit of the more tangible things. The important but not urgent, that it is so hard and yet so crucial to make time for. It is easy to push this onto heuristics, to decide by the numbers when you should do this, when you should do that, or redirect it onto your own manager and let them tell you what to do, what their strategy is. I believe that is a mistake that holds people back. You may feel that you don't have time to be strategic because there is so much else to do—so many meetings, so many pressing issues to address. But when time is limited, it is all the more important to use the time we do have well. We can't afford not to be strategic.

REFLECTION

Remember that good strategy is not the grand vision—it is real, tangible, proximate. Start with two questions:

- What is the underlying cause of the issue that currently manifests most clearly?
- What will be the next problem after the biggest problem you have today?

The first two things you need to understand to develop a strategy are (1) where you are at currently and (2) where you need to get to. You have the team you have today, your assessment of which should answer (1), and once you've obtained (either given or mined) enough direction, you should be able to have a sense of where that team needs to be in 6–12 months or longer, answering (2).

We'll talk about developing a strategy for what you need to deliver in Section 4, "Self-Improving Teams". In this section, we're talking about determining the strategy for how the team will evolve based on where you are today and where you need to get to. The two aspects of this are people and scope:

People
> Is the team expected to shrink, stay roughly the same, or grow?

Scope
> Is the scope of the team going to get smaller, stay roughly the same, or grow?

The three options each for both people and scope give us nine options overall, outlined in Table 6-1. You can read a full description of each area in Appendix A, but here we'll include three examples: layoffs, stability, and consolidation.

Table 6-1. Changes in people and scope result in shifting focuses and results

		People		
		↓	=	↑
Scope	↓	Deprecation	Reorganization	Consolidation
	=	Efficiencies	Stability	Growth
	↑	Layoffs	Expansion	High growth!

↓PEOPLE && ↑SCOPE: LAYOFFS

When there's more to do and dramatically fewer people to get it done, that's typically the result of some organizational shift going beyond efficiencies, such as layoffs.

What to consider

If your team structure has changed this much, you may need to rebuild your focus and processes from the ground up. This is not a time of incremental change but of dramatic rethinking. In this kind of situation, it may help to work on the assumption that things will get worse before they get better and operate accordingly.

What to worry about

Layoffs typically begin with the first round of attrition—the people who are let go—who, if it's done very top down, may take core knowledge with them. Then there is a second round of attrition, where the people who are left behind are significantly more likely to leave for other jobs, mixed in with the general disruption and destruction of morale such situations bring.[4]

When this level of change happens, nothing is constant, and that makes it much easier for people to leave. The first-order effects of a layoff may send people into challenging but possible situations, but the second-order impact can make those situations feel impossible. Hiring postlayoff is extra challenging as the perception will be that when there has been one layoff, there's likely to be another, and new people run the risk of "last in, first out." In countries with strong employment protections (such as much of Europe), if a company has laid off some class of worker, they may not be allowed to rehire in that group. While companies factor these things into layoff decisions, it's one thing for some VP to decide that they only need 20 site reliability engineers (SREs) versus 30, and it's quite another for the person who is trying to staff a 24-7 call rotation with 15 people a couple of months later.

What to do

First, look for your own oxygen mask. These situations are horrible; the day I had to lay off half my team was the single worst day of my professional career. Find someone to talk through your feelings and any survivorship guilt with—you're allowed to feel how you feel.

Look at your current scope and the team you have:

- Review your business-as-usual (BAU) tasks and consider how (if) they work with a smaller team. This is the work that can easily drown a team that is suddenly much smaller than it used to be, and you will need to figure out how not to keep everything from grinding to a halt.

---

4 The *Harvard Business Review* article "What Companies Still Get Wrong About Layoffs" (*https://oreil.ly/ bIB2K*) outlines the landscape around layoffs and some of the negative impacts that arise, including when internal and external communications are simultaneous and result in a bigger outcry, undermining trust with remaining employees. The data shows that layoffs rarely improve profitability, and companies that do layoffs tend to underperform for nearly three years postlayoff.

- Look at your current project work and aggressively reprioritize it. What needs to be cut?
- Identify the people who are most critical to retain, and see what you can do to make that more likely. Make a backup plan in case this fails.

=PEOPLE && =SCOPE: STABILITY

In the venture capital (VC) tech model, there's often this mentality of "growth or death," but that is not as true elsewhere. Some companies successfully bootstrap, while some nontech companies have stable tech organizations serving a growing nontech business. Sometimes, companies look to sustainable operations to ride out a rough period, like a recession, and then invest in growth again once the market recovers. Even growth companies may have parts of the business that operate sustainably. Often in tech, particularly during the boom times, we have operated as though growth in people (whether employees or users) is synonymous with growth in revenue, but that is not necessarily the case.

What to consider

Stability can be an opportunity for depth and refinement. In high-growth periods, we're always playing catch-up, but when things are stable, we can catch our breath and work on some of the nice-to-haves.

What to worry about

It's easy for stability to move to a place of efficiency—for instance, when someone leaves and there's no backfill. Don't be complacent; assess the risks and manage them.

This space can feel boring and lacking in opportunity for people who are focused on growth, and you'll need to get creative about how you give it to them—or accept that they will find it elsewhere.

What to do

If you're the ambitious, high-achieving kind, it may feel weird to be in a situation where things are going to be a little less exciting and stressful. If this is the case, you will first need to find a way to enjoy the time when things are a little more peaceful.

Then:

- Identify any areas of complacency where things are not really up to standard but haven't (yet) caused problems. For instance, if you are an internal-facing team without monitoring, relying on people to ping you on Slack, you may want to put in a proper alerting system.

- Identify any areas that rely on this stability to work. What would break if there were a significant change? You may not need to address this now, but it's worth knowing about and having a plan. For example, if you have a very manual release process, you may want to know what it would take to automate it.

- Identify if this stable state is holding anyone back from advancement, and consider what is available to give them (if anything).

↑PEOPLE && ↓SCOPE: CONSOLIDATION

Early-stage companies often go through a slightly chaotic process of trying to find product-market fit; once they have found it, there's this period of making everything work the way that users (outside the early-adopter space and/or paying) expect it to. At this point, responsibilities that were grouped together get split apart. The feature that was maintained by one person is now maintained by four; the app that two people built in React Native is now two native apps and two separate teams of six people.

What to consider

This stage brings with it a lot of growing pains as it may be part of transitioning from individual work to a team-based structure where people collaborate more and move from being generalists to specialists. It can create dissonance—especially for founders—when overall a lot more people are present, but the organization seems to achieve less than it did with fewer people. It's also a stage when early technical decisions may come back to bite you.

In this phase, you need the old guard because they know how everything works, but how things are done needs to change because they no longer scale; the old guard may or may not get on board with that. If the old guard can, and are willing to, grow with the team, that's great. If not, they get overtaken by new hires, who may be more specialized or just have more overall experience, and they may get frustrated and leave—or worse, get frustrated and stay.

What to worry about

The first thing is bringing people with you as much as possible through the change—getting them to understand, accept, and adapt to it—but know that is not always possible.

The second is scaling development. There are different requirements when many people are working close to one another in a codebase for developer experience and release processes, and these things need to evolve as your team does.

What to do

- Start extracting tribal knowledge and getting it persisted in a written form.
- Pay attention to and start addressing the pain points of your development process:
    — What processes need to be automated?
    — What parts of the code base are disproportionately painful: more expensive to change and harder for new people to ramp up on?
- Update processes for a larger team; we'll talk more about updating process in Section 4, "Self-Improving Teams".
- Build your understanding of the old guard. What do they know that no one else does? What lives in their head without being written down? What are their skills?

Developing and iterating on strategy

Your strategy is how you expect to close the gap between where you are and where you need to be. But a strategy is a living thing that needs to adapt to changes: if the current situation shifts, you'll need to recalculate. If the end state changes, you'll need to adjust. If you *meet* the end state, you'll need to start again from the beginning. And when the expectations of your strategy meet the reality, you may discover that you need to change your approach.

My boss pushed me to hire some additional leaders for my team for a while before I did it. He had more clarity about the level of growth we were expecting to achieve and how much more leadership I could really wring out of the team as it was. I was too slow to come around to this, for two reasons. The first was that I was already maxed out and didn't want to deal with building out a leadership hiring process in addition to everything else. The second reason was that I was too attached to my "grow from within" playbook that I'd honed in my previous role, where hiring people into leadership roles wasn't an option. I had built a strategy that worked to grow the team from 8 to 30, but it was time to change to a different strategy, which crucially included hiring in some help! This new strategy would need to turn the now ~30+ person team into three separate teams of 10–25 each...all with the additional challenges of navigating a "flat organization" with no concept of groups of teams. I had to let my original strategy go to allow it to evolve and grow.

Come up with your strategy. Write it down. Think about the constraints you'll need to break and how to accomplish that to get to your desired end state. But don't fall in love with your strategy. Don't be blinded to its limitations. Strategy is like a pair of shoes; it needs to be suited to the terrain—and thrown out and replaced periodically. Your favorite road-running shoes will not survive long on a snowy mountain, and your muddy hiking boots will be frowned upon in a fancy city restaurant. The more options you have, the more places you'll be able to go, and next we'll talk about one of the most extreme options: that of the struggling team.

APPROACHING STRUGGLING TEAMS

When looking for new challenges, taking on a struggling team can be both a temptation and a risk. Struggling teams are places of opportunity, both for leaders and other team members, to demonstrate dramatic improvement and impact. Often when a team is struggling, it's because the team didn't navigate the organizational shift of scope and people in the way that was required: the scope is off, and people are a mix of over- and underutilized. Having a sense of the strategy required—where the team is at and where it needs to go—can be the most useful input in deciding whether to take on such a team.

But we shouldn't gloss over the very real risk that the team will not improve and may ultimately be disbanded, presenting a career risk for everyone involved. Nor should we gloss over the fact that bias and inequity mean that certain portions of society are much more able to take career risks than others, although sometimes the riskier proposition is the opportunity—a phenomenon known as the "glass cliff" (https://oreil.ly/ZqvZf).

Successfully navigating the turnaround of a struggling team means doing the hard work of change management, which is a daunting task. The first question is: should I take this on? Let's walk through three aspects of this decision process.

Step 1: List your biggest problems

Struggling teams have a list of common symptoms, with failure to deliver usually being the first big external diagnosis. But delivery is a trailing indicator. Looking more closely, you'll often see a litany of things that *cause* the failure to deliver. Attrition. Poor or missing roadmap. Lack of ownership. Lack of necessary people or resources. Over time, you will want to address all these things (and we'll talk more about some of them in Section 4, "Self-Improving Teams"), but step 1 is to identify the biggest bottlenecks that are currently holding the team back.

Example 1:

> The team in question is tasked with delivering user-facing features but doesn't have a designer.
>
> This team needs one or two full-time designers to deliver on the stated priorities. Will you get the staffing you need to make the team successful?

Example 2:

> This team has four ongoing projects. The status of each is unclear, although it is agreed that all of them are running longer than expected.
>
> The first thing to do is to define what a shippable state is for each project and assess where the team is in relation to that. Going forward, you will need to institute better up-front planning, clear deliverables, and metrics for all projects. Are these achievable goals?

Step 2: Decide how to debug the cycles

All teams fail, at least on occasion. They miss dates, lose people, or get caught by surprise. All teams fail because they are made up of humans, and failure is the human condition. The distinction of high-performing teams is not that they never fail; it's that they fail infrequently, learn from those failures, recover quickly, and fail differently over time. Failing teams, meanwhile, snowball into a state where failures in one place create failures in other places. This cascading effect compounds the problems, making them harder and harder to fix.

Example 1:

> The team is understaffed because of attrition. This is largely related to poor promotion rates, which are a function of poor delivery (and a promotion process that emphasizes shipping new features over anything else). Meanwhile, the reputation of the team is low, and this makes it hard to attract people into it.

Can you legitimately give people a reason to stay? Can you lay out a road-map that demonstrates impactful, promotable work and ties the work of individuals to it? Can you summon support from senior leadership in the next promotion cycle and in recruiting?

Example 2:

The team is not delivering on key projects, nor are bugs being fixed. This is leading to a perception among leadership that the team can't deliver at all, resulting in a smaller and smaller mandate. The team, meanwhile, resents being distracted from project work by all the bugs and argues this is slowing them down.

You need to clearly delineate maintenance work and start delivering incremental improvements that help break the cycle. You will have to work to improve trust and transparency outside the team while filtering the criticism to make it understandable, fair, and actionable to the team.

When teams are defensive, they can construct a narrative together that deflects the external criticism—and they probably talk to one another a lot more than they get external input, making this narrative hard to break. You need to change the narrative from "This team isn't delivering" to something more concrete and actionable, like, "This key project is delayed by six months for reasons that we really should have understood up front; meanwhile, our app store reviews are increasingly critical and generating a high volume of requests for support because we went 'all hands on deck' for the key project and haven't addressed consumer feedback on anything else at all."

Step 3: Define the constraints and understand the latitude you have

Getting through a turnaround requires support from your direct boss and peers. You need to understand what they expect, what they will support you with, and what latitude you have. Can you focus first on staffing issues and let delivery improvements slide for a while? Or do you need to focus on delivery and work with what you have? Does leadership have a very different sense of the problems the team is having? And are you able to make the higher-level view gel with your more detailed one?

It's important at this point to ask for what you need to be successful, and it's better if you are in a strong negotiating position to do so. For instance, when a manager at a previous job gave me six months to turn around a team, one of my requirements was that, if necessary, I would be able to send struggling individuals to other teams. These kinds of nonnegotiables are part of being up-front about the extent of the work you can and can't commit to in the given time frame.

Now what?

Once you have agreed on the problems, you get to make a plan. There is always more to do than we have time for, but with a struggling team and many things to fix, this is especially true. The key thing is to identify the highest-leverage activities, achievable by the team in the current state, so you can address your biggest problems and focus your energy on them. As part of this process, you also may identify useful things to do later—but not yet.

Perhaps your list looks something like this:

1. Clarify the roadmap for the next three months by the end of next week. A longer-term view will also be needed, but that can wait a couple of weeks.

2. Kill a failing project and reallocate staff to other priorities.

3. Replace the manager of the operations team with someone who is more focused on delivery and has better interpersonal skills. (Here you might list some internal candidates who are potentially a good fit.)[5]

Item zero, 1.5, 2.5, and 3.5 on your list is to listen. Struggling teams are made up of struggling individuals who need to be heard. They are surrounded by peers and partners who feel let down. Listening can be scary because it often involves admitting that the problems others are experiencing are not the problems you're taking on right now. But if you are clear about what you're working to address and why, and you can lay out a path for improvement, people are more likely to give you latitude than if you ignore or avoid them. Then, if you can demonstrate that you're meeting the milestones you laid out, you build trust.

The other thing to recognize is that this process is often not a one-time thing but rather a cycle. As you tackle the first set of problems, new ones emerge. Midway through my first turnaround project, I produced a 13-page document—meant for me, not for wide consumption—that outlined the constraints we had broken through and the constraints that were coming next. It was an emotionally draining but fascinating and cathartic process. One of the things I found most

---

5 This specific action might be hard or may take a while to move forward, depending on HR or legal constraints, but don't let those stop you from outlining what you *want* to do—ruling out options too quickly constrains your thinking.

interesting was how the same problems surfaced, in different forms, again and again. We would break constraints on, for example, communication or project planning, but as other areas improved, those things would become constraints again.

So, should you take on the turnaround project? That depends on whether you believe you can succeed—and part of that is the level of support you will have to do so. These situations invariably feel like "taking one for the team," and when that's valued and supported, it's great. But when it's not...it sucks. I'd encourage you to make a rational assessment and have the conversations with your boss you need to in order to feel confident taking it on, rather than deciding based on emotional responses to commitments or risk. Working through this process and developing a strategy will help you make a deliberate decision—and will set you up for success.

Letting Go: Why Your Job Should Change Regularly

Leadership roles evolve, especially through periods of transition. As a leader, I have found my own role changing as challenges on the team change. During high-growth periods, around every four to six months in I realize everything is fundamentally different and the way I need to spend my time changes, too.

When the team changes dramatically, it's worth asking the people most affected by your work some questions:

- What do you see as the most important thing(s) I do (generally)?
- What are the most impactful things I do for you specifically?
- What is one thing you think I should stop doing?
- What is the biggest area of your work where you want/need me to support you?

But even when it's not as obvious, your job as a manager can still evolve. Maybe you used to have mostly new managers reporting to you, and now they've found their feet, meaning you can be less involved and spend your time on something else instead. Maybe your team had some kind of pressing problem—a big project with a looming deadline or bad releases that needed to be fixed—but now it's on track, so what do you focus on next?

I would go as far as to say that once you have significant responsibility, your job should change at least every six months. Your team should be evolving, and you should be leveling up your direct reports. So at least twice a year, set aside some time and think about the following questions:

- What are the biggest challenges for my team?
- What are the biggest challenges for my peer group?
- What are our biggest challenges as an organization?

Then look at how you spend your time. Does it align with the challenges you've laid out? Does it reflect your priorities?

Ask:

- What can I delegate?
- What should be dropped?
- What new things do I need to take on?

These can be difficult questions to ask when you're overwhelmed, and the answers we land on may scare us. Having a coach can be useful here as it forces you to step back, consider the big picture, and explain it to someone who is not deeply involved nor invested in any outcome but you being your best self.

Left to our own devices, we might avoid doing the work of figuring out this evolution, either because we don't want to confront the current challenges on our teams, or because we're afraid to give things up, or because we're so maxed out that we can't contemplate the thought of assigning ourselves even one more thing to do. But these thoughts are illogical and counterproductive at best, and downright destructive if not kept in check. So let's dismantle these ideas one by one.

YOU DON'T WANT TO CONFRONT THE NEW CHALLENGES FACED BY YOUR TEAM?

This is literally your job as a manager. It might be hard, but you need to accept that and recruit the support you need from your manager or broader network. These situations rarely get easier as time passes, so if you fail to do it, you'll eventually get found out.

It can be hard to admit we're afraid of something, and we all avoid things sometimes—especially if we're feeling overwhelmed. Try to give yourself some space to think, bring the topic of "next team challenges" to your 1:1 with your manager or your coaching call, or try a free-writing exercise (*https://oreil.ly/RPtOZ*).

YOU'RE AFRAID TO GIVE THINGS UP?

Perhaps you fear you won't have enough to do—but I can promise this is almost certainly not going to be the case.

Perhaps you fear a loss of status or information, believe that your value is measured only by your direct contributions, or worry that people won't do things to your standards or that if you are not in certain meetings, you will be seen as unimportant or unnecessary. Keep in mind, managers who can successfully replace themselves are extremely valuable. If you can't trust your organization will value this, then you have other problems; just know it's a skill that will be valued elsewhere if not where you currently are.

It's also worth remembering that while meetings are often part of your job, it's no one's job to go to meetings. What are meetings supposed to accomplish? Does the meeting need to exist at all? Can the outcome be accomplished more efficiently? Do you genuinely add value? Will anyone or anything be worse off if you took the meeting off your calendar or gave your place at the table to someone else?

YOU'RE AFRAID TO TAKE ON SOMETHING NEW?

When you feel so overwhelmed that you cannot do any more, it's easy to believe there's no time to step back, no time to invest in other people to take some of the workload. This is madness: you cannot overwork yourself out of the overwork trap. The only way out is to work smarter (strategically!) and to get rid of some of the stuff that is currently filling your days, whether that means someone else does it or it doesn't get done at all.

So, where to begin?

First, think about the team and what kind of strategy you need to be successful:

- List your team challenges: what are the most pressing things for your team to accomplish over the next quarter or year?

- List your team constraints: what are the biggest things currently holding your team back? What constraints, if broken, would unlock the most potential?

- What's the gap between where you should be and where you are? How can you push that forward?

- Build peer support into your team practices rather than expecting to be the person who supports everyone else. Escape the trap of needing to be useful!

- Look at the big picture at your organization and consider how your work and your team fit into it.

Then think about yourself and review how you manage your energy:

- Rebuild your schedule: eliminate everything from it and start from the beginning. What goes back in?

- Ask people what you do that is most valuable.

- Give away things you know.

- Give away things you're not sure about.

- Ask your peers what they need.

- Ask your boss what they are worried about.

Finally, take a vacation. There's nothing like having people manage without you for a bit to highlight the things you can let go and the things you do that are really useful.

Remember, you have value, independent of your economic utility. Your wants and needs matter. It's okay to keep something you enjoy as long as it's not getting in the way of other people or other priorities. We don't have to make our teams happy at our own expense. Fulfillment—like everything else—is a team sport. And when you're confident that the new version of your role will be at least as fulfilling as the current iteration, it's that much easier to embrace the change.

7

Expanding Your Leadership Range

In leadership, as in everything, we all have our defaults: the things we turn to first or just go ahead and *do* without thinking. Think about the rest of your life. Do you have places you like to go on vacation? The first machine you head toward at the gym? The genre of book or movie or music you turn to when you've had a bad day? The restaurants you think of first when planning a night out? Probably. But even if you don't, even if you turn to novelty each and every time, *that too is a habit.*

People are relatively predictable. Which makes sense—anything else would be exhausting. The origin of bias, of stereotyping, is our brains taking shortcuts for the purpose of efficiency. But taken too far or applied poorly, this same pattern of thinking becomes harmful, which is why we have to reprogram ourselves to think differently.[1]

We're typically not different people at work than we are in the rest of our lives. Our leadership range reflects who we are, what we value, how we like to show up, how we want to be seen. Like everything, that probably works better in some situations than in others, and if we want things to go differently, we have to change the way we approach them.

In this chapter, we'll work through identifying your failure modes (which are often overused strengths) and setting up warnings so that you can self-correct. We'll look at the different leadership styles, when they are effective, and how to expand your range.

---

1 This idea of "fast and frugal" thinking was coined by the psychiatrist Gerd Gigerenzer and is covered in Malcolm Gladwell's book *Blink: The Power of Thinking Without Thinking* (Little, Brown). For more on how we are affected by stereotypes, I recommend *Whistling Vivaldi: How Stereotypes Affect Us and What We Can Do* by Claude Steele (W. W. Norton).

Identifying Failure Modes

If I think honestly about the ways in which I've failed, they fall into some categories:

- Getting too overwhelmed
- Avoiding or deferring decisions because I didn't want to deal with people's emotions
- Not trusting my judgment around things that seemed "off"
- Letting people ignore the severity of something—communicating something (often up or sideways) but not making the person *listen*
- Taking too much "for the team," both in volume and in specific activities that I got no value from, because someone needed to do them

There are probably more, but that list seems sufficient to put in print. When I look at those failure modes, I see that they come from values around being driven and achieving, caring about people, not being judgmental, allowing people to make their own decisions, and being a team player—all things I like about myself and want to hold on to.

REFLECTION

What are your values? How do they show up as your failure modes?

Early in your career, there are things that you obviously just need to get better at. But as your career progresses, the ways in which you get in your own way are more often the flip side of the things you are great at.[2]

We've all seen this play out at work. The great strategist who can't execute. The single-minded visionary who misses the bigger picture. The exacting leader with high standards who cultivates both great products and an abusive work environment. Some people are so attached to the narrative that sometimes they claim the weakness *as* the strength. Say that it's not possible to have high standards without also being abusive. That in order to be successful we have to eliminate every other aspect of our lives.

---

2 *What Got You Here Won't Get You There* by Marshall Goldsmith (Profile) covers this in depth.

I don't believe this has to be true. Two tools can help: the CliftonStrengths assessment and Positive Intelligence.

We've all heard the saying "When all you have is a hammer, everything is a nail." The purpose of identifying our failure modes is not to deny our strengths but to understand when our strengths can be usefully applied and when they cannot. The idea is to *maximize* our strengths while sidestepping as much as possible our controlling weaknesses.[3] Sometimes we will need to address our weaknesses—when they are "controlling"—and sometimes we will need a partner to address them, such as when we hire to fill them or even build a "co-leadership" structure.[4] We'll talk more about hiring in Section 3, "Scaling Teams", but having a more holistic view of strengths and letting go of judgment relating to their value will help us avoid building monocultures in our teams.

CLIFTONSTRENGTHS ASSESSMENT

The CliftonStrengths assessment is a 30-minute assessment from Gallup that identifies your natural inclinations and maps them to the 34 CliftonStrengths themes (*https://oreil.ly/vvZEO*) in four categories (Table 7-1). Our "strengths" are our top five, each one of which has corresponding weaknesses—failure modes—that can occur when they are overused.

Table 7-1. CliftonStrengths themes

Strategic thinking	Relationship building	Influencing	Executing
Analytical	Adaptability	Activator	Achiever
Context	Connectedness	Command	Arranger
Futuristic	Developer	Communication	Belief
Ideation	Empathy	Competition	Consistency
Input	Harmony	Maximizer	Deliberative
Intellection	Includer	Self-assurance	Discipline
Learner	Individualization	Significance	Focus
Strategic	Positivity	Woo	Responsibility
	Relator		Restorative

Understanding what comes more naturally to you can help you have more time and patience for people who have different things come more naturally to them. Maybe you are naturally strategic, which is great, but someone who is not

---

3 The concept of a *controlling weakness* comes from Marcus Buckingham's *First, Break All the Rules* (Gallup Press) and is defined as a weakness that is limiting and holds us back from being effective or advancing.

4 *Co-leadership*: a pair of people leading together, ideally leveraging each other's strengths.

naturally inclined to strategy will need to work harder and be more deliberate about taking time to develop strategy than you will. If that person is better at influencing others—for instance, through an activator strength—then by working together you can both increase your impact.

If you are extremely analytical, which is common in the tech industry, the associated weakness tends to be overthinking or overanalyzing things, which can lead to something of an analysis paralysis. Knowing this can help you notice when it happens and set up strategies to extract yourself from it.[5]

REFLECTION

How do your strengths show up in your current job?
What strengths do you take for granted?
What weaknesses are the flip side of one of your strengths?

POSITIVE INTELLIGENCE

Positive Intelligence, or PQ, was founded by Shirzad Chamine, former CEO of CTI (*https://oreil.ly/dbMUF*) (now Co-Active Training Institute), a coaching training and leadership development organization.[6] In coaching, we often talk about "saboteurs," which you might also describe as the darker voices in your head or the ways that you get in your own way.

The foundation of PQ is the factor analysis that reveals the core saboteurs most people have some combination of. Following is a list of saboteurs; visit "How We Self Sabotage," Positive Intelligence (*https://oreil.ly/gWkiP*) for detailed descriptions:

- The Judge
- Avoider
- Controller
- Hyperachiever
- Hyperrational

- Hypervigilant
- Pleaser
- Restless
- Stickler
- Victim

---

5 I hate booking my own travel because I start treating it like an optimization problem and overthinking everything. The only way I managed to be nomadic for three years was by having a guy who did all my travel booking for me; otherwise, I would probably still be trying to figure out how to get to my first destination.

6 Disclaimer: I did my coaching training through this organization.

At the core of our relationship with our saboteurs is the belief that *our saboteurs keep us safe*. My top saboteurs are, as my coach put it, "all the hypers," and as much as I can recognize that all these patterns of thinking have made me exhausted and miserable, I also worry that without them I would just...lie down and do nothing for six months.

There are two things I really like about the PQ saboteur model: (1) It's efficient. You take a five-minute test and get an answer. (2) It's impersonal and separates the saboteur from the personality. We can then talk about the "victim" or the "controller" as separate from the person, something they can choose whether or not to listen to.

Try taking the test (*https://oreil.ly/LOo9J*) and see what you learn about yourself.

REFLECTION

What situations do you notice your saboteurs most present in?
What do you think they are trying to protect you from?
How do you think they are getting in your way?

Communication Failure Modes

There's a saying in software that all bugs are eventually user interface bugs, because someone has to see them in order to report them. At work, it often seems like all problems are eventually communication problems, because communication is the way we interface with one another—and the way most problems surface. Some communication failure modes to watch out for include:

- Lack of depth

- Conflicting context

- Too much empathy

- Missing empathy

- Assuming unearned trust

LACK OF DEPTH

Lack of depth shows up most when communicating strategy: if you are trying to lay out the way forward, you need to be able to outline what is actually changing. If you stall on concrete questions, this will create confusion and frustration as people are unable to connect the changes to their day-to-day responsibilities.

Example 1:

> Running a workshop without a clear goal and outcome. People will not understand what they are doing and feel like their time was wasted.

Example 2:

> Outlining plans for a product improvement but glossing over the underlying work that will be needed to support it—and the implications for support staff.

Good strategy requires depth, which means having an understanding of key problems and an ability to explain how the strategy addresses them. When that depth is absent, credibility is lost, and the strategy often ends up being ignored.

CONFLICTING CONTEXT

Conflicting context shows up when communicating superficially, without considering the context held by the other party.

Example 1:

> Team A requires support from Team B on its top-priority project. But Team B considers this project low priority and drops it in favor of something else, leading to frustration from Team A.

Example 2:

> Person A and Person B have been given overlapping mandates. Both are trying to deliver, but their priorities are in conflict.

Once, when a team experienced a high-profile cascading failure, a colleague and I spent five hours with them doing a detailed retrospective over the full timeline of the project. We found that the communication disconnect happened *months* before the eventual failure; at some point, the context of this team and the people they were working with wildly diverged, and while work and communication continued, they were not talking about the same thing anymore.

People operate in different contexts, and what's obvious to you may not be obvious to them—and vice versa. When you have a lot of shared context with someone, much information is implied or already exists between you, and you don't need to restate it. However, with people who exist in a different context, you may need to provide more background and information and ask more questions; otherwise, you risk talking past one another.

TOO MUCH EMPATHY

Communicating with too much empathy may mean you spend so much time on people's feelings that you never get to the point.

Example 1:

A manager needs to make a change on the team but knows someone who will be upset by it. So much time is spent trying to prepare them for it and manage their emotions that the change gets delayed.

Example 2:

The team is having unproductive arguments, but the manager doesn't want to make people feel shut down so doesn't intervene. Meanwhile, the quieter people on the team are upset by the arguing and get even quieter.

MISSING EMPATHY

Empathy shortages are arguably a chronic condition at many companies, but they're particularly problematic when you're communicating important organizational change.

Example 1:

External factors to the business cause leadership to review the roadmap and make the difficult decision to end a project and reallocate people to higher-priority projects. When communicating this, the focus is on the business need, with no acknowledgement of the investment people have personally in a project they've been working on for a long time, creating anxiety and resentment.

Example 2:

An aggressive hiring goal makes a reorganization necessary to support the increased size of the team. This means that some teams will initially be smaller. Some leaders will feel like they had responsibility taken from them, and some individual contributors are surprised by a shift to a new manager. Everyone panics.

An organizational change is usually something that leadership has spent a lot of time evaluating and working through, with the broad insight to see why it's necessary. By the time the plan gets communicated more widely, senior leaders have worked through their own personal feelings about it (if they had any), perhaps forgetting that for many people now hearing about the plan, this is just the beginning, and they will have their own emotions and questions to work through.

There's often a similar dynamic in feedback conversations, when the person giving the feedback is caught up in their own emotions about it and forgets about the feelings of the person receiving the feedback. When communicating anything that people have the potential to find difficult, it's crucial to have empathy and create space for people to process.

ASSUMING UNEARNED TRUST

This shows up when people expect or assume more trust than they have.

Example 1:

An engineer proposes a dramatic structural change to some code and meets unexpected resistance from the team. The team, it turns out, is harboring residual resentment from the last big change the engineer recommended, which caused a number of problems because its complexity was underestimated.

Example 2:

A new hire to a leadership role proposes a major change without explanation and is confused when they're met with skepticism, even when the new leader has a track record of successful reform elsewhere. People aren't interested in that track record, though; they see their company culture as unique and presume that what worked somewhere else won't just work here.

As a manager, it's important to give trust (along with clear expectations and support) to earn trust. As a person new to a role or company, it's important to remember that trust always needs to be earned from at least some people beyond what was assessed in the hiring process, so it's safest to operate like you need to earn it from everyone.

Growing Your Impact

There are a thousand different job-leveling frameworks out there, but I believe the core of growth in an engineering context comprises four things:[7]

- Scope
- Complexity
- Output
- Agenda

When working within a job-ladder framework, it can be a challenge to (1) separate evaluative feedback from developmental feedback and (2) truly focus on growth rather than trying to check a box. As such, I hope this framework can be a neutral tool you can use to consider how you can grow and what kind of growth would improve your effectiveness in your current role and, longer term, your actual career goals. It's fine to enjoy scope more than complexity (and vice versa); the main thing is not to allow something to hold you back by becoming a controlling weakness. Each one of these requires a shift in mindset as you progress, which can be the hardest part.

SCOPE

Scope is the size of the problem or responsibility and is relatively straightforward to quantify in volume and type of changes to the codebase or people involved.

For ICs, this often goes bug → small feature → large feature → multifeature project →...

For managers, it's often TLM (tech lead manager) for small team → manage team → manage managers → manage managers of managers...

Mindset shift

When we grow in scope, we need to learn to let go of control. What is possible to control in a small team or project gets progressively harder over time. Everyone reaches some kind of limit there (admittedly, some much later than others).

As scope increases, you will need to figure out how to get the outcomes you are responsible for with different mechanisms than that of control (documentation, setting standards, coaching and mentoring, reporting, proactive check-ins, etc.).

---

7 This model came out of work with a coaching client, Akshay Shah.

Opportunity

As scope changes, think about how the way in which you work needs to change. While we might fix a bug using a debugger, it's much harder to develop a complex feature like that, and so that increase in scope is an opportunity to adapt and grow (and start writing tests). This continues as you take on more responsibility; you cannot (effectively) manage managers the way you manage ICs.

It's worth noting that sometimes in our careers we might take a decrease in scope—for example, when we change jobs and have to onboard through fixing bugs or go from managing managers to managing ICs in a smaller organization. In these instances, we will also have to adjust our approach to our new scope in order to be effective.

COMPLEXITY

Complexity is the difficulty of the problem and is more difficult to quantify. Sometimes it takes some amount of domain knowledge to know what is challenging.

For ICs, complexity is in the projects, the complexity of the feature, the problem space, or the codebase.

For managers, complexity is in what is being delivered and how:

- What is being delivered is the complexity of the product: managing a team building a CRUD app is less complex than managing a team building a more specialized or unique application with a greater variety of client-side work.

- How it is being delivered is the way in which the team works: for instance, a team that has to manage more stakeholders (such as a platform team) may have higher complexity than a product team.

Mindset shift

Complexity cannot continuously increase indefinitely. Effective scaling (of an individual or a team) involves the transformation of complexity, where something that was complex is made simpler in such a way that it can be more sharded.

Complexity is often a point where engineers around the senior level get stuck. Having worked with continuously increasing complexity, they either make things more complex than they need to be or persist a high level of complexity that needs to be simplified and sharded.

Opportunity

Take a holistic view of complexity: complexity is not just the code but also the product and environment it exists within. Focus on outcomes to resolve, and reduce complexity in such a way that allows others to be successful. For instance, you might modularize a complex application or abstract something complex so that other people can work with it more effectively, without needing to understand the internals.

OUTPUT

Output is getting things done or causing things to get done.

For ICs, this starts with an initial focus on raw individual output and progressively shifts to enabling other people over time.

For managers, this is the output of your team or org over time. Over time is important: while you can create short-term output at the expense of other things (underlying work, attrition), you will typically pay for those in the form of lower output later. This means being able to balance an increasing amount of disparate work while maintaining a good rate of output consistently.

Mindset shift

To improve output, you have to think progressively longer term over time. Individuals who focus on their own effectiveness at the expense of the team take too short-term a view and need to reorient themselves around making the *team* more effective. Similarly, managers who push short-term output at the expense of longer-term team functioning need to think more about driving output in a way that is sustainable over time.

It's critical to recognize when output needs to happen through other people rather than individually and understand how to balance short-term pressures with longer-term impact.

Opportunity

Take a broader view of output: identify where you can increase output outside your own individual efforts (through other people and/or via processes, tooling, architecture, etc.).

AGENDA

Agenda is the driving force of what you're trying to get done. As you progress through an organization, you will have to balance a broader agenda, ultimately driving the agenda of the entire organization and navigating conflicts between that and the agendas of individuals or teams.

ICs often focus on their own work or team, but at Staff and above, they typically need to hold a larger agenda around an area or domain.

Managers need to fit the agenda of the team within the overall agenda of an organization, tying it to work that impacts the business.

Mindset shift

When people take too narrow an agenda that conflicts with bigger agendas, such as the IC who pursues their own interests at the expense of the overall team or a manager who is too focused on the team they manage versus taking a "first team" (peer) mindset, they focus on the immediate vicinity at the expense of what's outside it. Take a step back and think about how your work fits into the context in which you operate—and beyond.

Notice moments of conflict between an individual or team agenda and an organizational agenda. This is an opportunity to surface agendas and resolve conflicting priorities in the context of the agenda of the organization.

Opportunity

Look for places where there is an opportunity to tap into a broader agenda, whether it's focusing on the team's needs rather than on your individual needs, or on the goals of an overall initiative rather than on the pressures of the project.

REFLECTION

Think about how these areas show up in your day-to-day work. Which one would it most benefit you to improve?

Understanding and Expanding Your Leadership Styles

"The best leaders master multiple leadership styles," blithely comments some post on leadership.[8] But how? Many leaders are overly reliant on a style, and this can hold them back. Generally, leadership styles are a function of emotional intelligence, and working on emotional intelligence, such as working on becoming more coachable (as we discussed in Chapter 3), may help, but how do we work on this aspect specifically?

That depends on the style you want to build out. In 2000, Daniel Goleman published research in the *Harvard Business Review* identifying six main styles of leadership (*https://oreil.ly/BIi9e*), each originating from different aspects of emotional intelligence (*https://oreil.ly/U4Xbu*): pacesetting, authoritative, affiliative, coaching, coercive, and democratic.[9] Even though our expectations of the workplace and of our managers have changed since then, these styles are still a useful place to start when considering what situations call for and the styles we tend to default to ourselves.

None of these styles is appropriate for every occasion, and some need to be used more selectively than others. But they are all worth being aware of, and the broader our range, the more we can consciously think about what leadership styles to deploy, rather than sticking to our defaults.

Let's go through the six leadership styles presented by Goleman. His article (*https://oreil.ly/BIi9e*) gives some solid explanations. Then we can think about how they are useful.

PACESETTING

Pacesetting leaders expect excellence and self-direction. A leader who sets high performance standards and exemplifies them himself has a very positive impact on employees who are self-motivated and highly competent. But other employees tend to feel overwhelmed by such a leader's demands for excellence—and to resent his tendency to take over a situation.

—DANIEL GOLEMAN, "LEADERSHIP THAT GETS RESULTS"

---

8 OK, I admit it—it's my post (*https://oreil.ly/SyGx9*).

9 Note this is paywalled.

When is it useful?

Pacesetting is for those times where there's a lot to be done. Maybe you're trying to ship something, digging out a backlog, or otherwise surviving an intense period. Direction should be clear, with minimal need for collaboration or experimentation.

What do you need?

Stamina and focus. The pacesetting leader is rarely known for their work-life balance. You need to be ruthless in saying no to the things that are not core to your effectiveness.

Situations to seek out?

Look for the numerically measurable problem, where everything is fixable by just Doing the Work. Just make sure it's not indefinite and that you have what you need to be successful.

Shift your mindset

High expectations of other people are key to the pacesetting leadership style and what distinguishes the pacesetter from the generally hard-working leader. Most of the hardest working leaders know they cannot expect the same level of dedication from their teammates; their commitment and work ethic are often what have propelled them up the ladder. However, to be a pacesetting leader, you need to demand excellence from your teammates, which means you need to have a clear idea of what excellence is and be prepared to set expectations accordingly. Some people may match the style without you saying anything; others may have to be told.

I suspect one of the reasons why pacesetting has a negative effect on teams is that pacesetters are overreliant on working harder and miss the point where they need to work smarter instead. Make sure that you deploy this style when working harder will make the difference—and switch to another style when it won't.

AUTHORITATIVE

Authoritative leaders mobilize people toward a vision. An authoritative leader takes a "Come with me" approach: she states the overall goal but gives people the freedom to choose their own means of achieving it. This style works especially well when a business is adrift. It is less effective when the leader is working with a team of experts who are more experienced than he is.

—DANIEL GOLEMAN, "LEADERSHIP THAT GETS RESULTS"

When is it useful?

The authoritative style is helpful when you have a new, or lost, team that needs a way forward.

What do you need?

Well-founded confidence in the domain and your own expertise, plus the ability to tell a compelling story about what the team is doing and why.

Situations to seek out?

Seek out a problem where you have a strong reputation and deep expertise, potentially a slight adjustment or repeat of something you have done before. This might seem a little dull—especially if you prefer novelty—but will allow you to be more definitive and work more quickly.

Shift your mindset

The differentiator of the authoritative leaders is confidence. They believe they know the way, and they will lay out the path to get there. To take on this style, you need to believe that you know how to address the situation. This doesn't mean not listening to other people, but it does mean fitting that information into your mental model and pushing things forward. This style is best adopted when it comes based on a reputation earned elsewhere, ideally nearby, so you need to be able to take pride and believe in your past accomplishments, too.

AFFILIATIVE

Affiliative leaders create emotional bonds and harmony. The hallmark of the affiliative leader is a "People come first" attitude. This style is particularly useful for building team harmony or increasing morale. But its exclusive focus on praise can allow poor performance to go uncorrected. Also, affiliative leaders rarely offer advice, which often leaves employees in a quandary.

—DANIEL GOLEMAN, "LEADERSHIP THAT GETS RESULTS"

When is it useful?

This style is incredibly useful when healing a broken team or bringing together a team that is exhausted (perhaps by the pacesetting style...).

What do you need?

Patience and empathy. You need to be willing to hear people out, give people space, and let some amount of chaos happen as the team evolves.

Situations to seek out?

Look for the team you believe in, that you can see has had a hard time as a result of outside forces. Maybe you take over from a bad leader or at the end of a difficult time (e.g., a team that has been scaling).

Shift your mindset

Affiliative leaders believe that the responsibility for team health and culture lies with leadership. It means taking a deep and personal responsibility for the culture of the team and working to create an environment where everyone can be successful. You will need to rank team goals after team health (and have the space to do so).

COACHING

> *Coaching leaders develop people for the future. This style focuses more on personal development than on immediate work-related tasks. It works well when employees are already aware of their weaknesses and want to improve, but not when they are resistant to changing their ways.*
>
> **—DANIEL GOLEMAN, "LEADERSHIP THAT GETS RESULTS"**

When is it useful?

This style is useful when there's been a lack of personal development—for example, people haven't been getting feedback—or when stretch assignments come without support. It can unlock a huge amount of capacity in your team or organization.

What do you need?

Patience and optimism.

Situations to seek out?

Look for situations where people haven't been set up to succeed but have done OK under the circumstances anyway. For instance, look for people who were reporting to someone who didn't make time for them. Pay attention to how coachable they are and how they view team responsibilities.

Shift your mindset

Coaching leaders believe in each individual. Shifting to this style means shifting your mindset and evaluating each individual on the team not against your expectations but against their best selves. Coaching leaders are a buffer who believe more in people who don't believe in themselves; this creates a balance against those who tend to overconfidence (and failing upward).

COERCIVE

> Coercive leaders demand immediate compliance. This "Do what I say" approach can be very effective in a turnaround situation, a natural disaster, or when working with problem employees. But in most situations, coercive leadership inhibits the organization's flexibility and dampens employees' motivation.

—DANIEL GOLEMAN, "LEADERSHIP THAT GETS RESULTS"

When is it useful?

As much as we don't like to admit it, at times, as leaders, our job is to tell people what to do. People who default to this style do it too much, but we all need to be willing to use it at times.

What do you need?

Conviction.

Situations to seek out?

Moments where you know you are right and someone else is wrong, and it's your job to tell them so.

Shift your mindset

Dig into that righteous anger and channel it, even if you don't express it fully—as my therapist told me, "Anger is a sign you need to stand up for yourself." You don't need to—and probably shouldn't—yell, but you do need to be assertive and firm.[10]

---

10 Personally, I am never more icily British than when I am consumed with rage; one time I deployed this style, and the conversation started, "A note on etiquette..."

DEMOCRATIC

Democratic leaders build consensus through participation. This style's impact on organizational climate is not as high as you might imagine. By giving workers a voice in decisions, democratic leaders build organizational flexibility and responsibility and help generate fresh ideas. But sometimes the price is endless meetings and confused employees who feel leaderless.

—DANIEL GOLEMAN, "LEADERSHIP THAT GETS RESULTS"

When is it useful?

There's a reason why many new leaders start with around 90 days of listening.[11] This style is incredibly useful for a new leader building credibility in a new organization as it ensures that everyone feels heard and maximizes buy-in. This style can be useful even when you are confident you have a good idea of what's going on, as you can diffuse resentment by showing you are listening and making people feel heard.

What do you need?

Patience. The leeway to invest time up front in exchange for increased buy-in later.

Situations to seek out?

Seek out situations where you know what you don't know and where building the knowledge to be effective is part of your remit so that you'll be given the time to do so.

Shift your mindset

The mindset of the democratic leader is that we will make better decisions together. They would sooner have the "best" 100% result of the collective than the 80% that the individual might create alone. They are willing to prioritize that buy-in consistently, even when there are pressures that might make a more expedient solution more appealing.

---

11 For a more nuanced take on that, the book *The First 90 Days: Proven Strategies for Getting Up to Speed Faster and Smarter* by Michael D. Watkins (Harvard Business Review Press) is helpful.

WHERE TO BEGIN?

It helps to start with the end in mind. The goal is some level of comfort with all of these styles, even if you tend to use some of them very sparingly.

The easiest way may be to build laterally, starting with what you're comfortable with. If you default to the affiliative style (as many new managers do), start adding the democratic style and making sure that decisions are made (you may need to throw a little authoritative style in to get decisions over the finish line!). Work on getting better at coaching people. These three styles are incredibly compatible with one another, and mastering them alone will take you a long way. Eventually, though, you will need to branch out and embrace the other styles.

If you need to make a more extreme change, either because you are reaching the limits of your effectiveness or in response to feedback, it may be harder.

If you fall on the "softer" side—affiliative, democratic, coaching—seek out or embrace a situation where combining the authoritative and pacesetting styles is warranted. It should be relatively timeboxed, so know that once you're out the other side, your other leadership strengths will allow you to deepen and continue your impact. Make sure that you have (well-founded) confidence and the support you need to be effective, like a coach and supportive friends to get real with, because you'll need to continually project that confidence to the team, even if you don't always feel it yourself.

If you fall on the "harder" side—coercive, pacesetting—try embracing the democratic or coaching styles more actively (they are less emotional than the affiliative style). Find some people whose potential you believe in and work on developing them. Seek out situations where you have influence rather than authority and lean into them—for example, working across the organization. You'll have to take a deep breath and accept that progress won't happen in the time frame you think it should, but then...does it ever, anyway?

Navigating Change

The thing about change management is that it involves a lot of invisible work that's hard to follow from the outside. The two things that people see in change management are the change (toward the end, if they are paying attention) and when it goes (sometimes horribly) wrong. Here are some things to keep in mind when navigating a period of change. For more depth on change management, the books *Switch: How to Change When Change Is Hard* by Chip Heath and Dan Heath (Crown) and *Managing Transitions: Making the Most of Change* by William Bridges and Susan Bridges (Nicholas Brealey Publishing) are helpful resources.

THERE ARE NO HEROES

When I was working at a failing startup, a new CEO came in, gave an hour-long presentation about himself, and handed out a book to everyone about some heroic individual bringing about company transformation.[12]

There was definitely change. Within six months, I had to fire half my team. Shortly after, I left, and not long after that, the entire company shut down. Although it was horrible, with some distance it was a learning experience. My teammates were very capable—all of us have gone on to better things—but I wonder if the CEO and the people he recruited right before it all came crashing down felt the same way.

Within companies, the best effort is team effort, and that is never more true than in times of transformation. Executives who breeze in during troubled times extolling their qualifications and claiming some kind of "visionary" status never seem to work out. When someone claims to be a hero, what people hear is that they don't need—or want—others on their side. People won't buy into your ideas just because you tell them to; they have to believe in you. And that requires a lot more work up front and listening rather than bragging.

HIGH PERFORMERS AND LOW PERFORMERS STRUGGLE MOST

This one took me some time to grok. Of course low performers struggle—they are (understandably!) afraid of what change means for them. However, high performers seem to struggle even more, and eventually I realized why: high performers have found a way to succeed in the system as is, so they tend to see the problem as other people needing to get it together and be effective, and they underestimate the impact of the structural issues that hold others back. As a result, change seems like unnecessary overhead that is liable to get in the way of their actual work.

Essentially, low performers need to know the "what"—what the expectations are in the new order of things—while high performers need the "why" of the change explained.

Before you try to introduce any kind of "performance management" to a team, the first step is always to bring in standards, support, and accountability. Once you have that, you can clearly communicate where people need to develop, give low performers the help they need and set them up to be successful, and if it still doesn't work out...let them go. This is not an easy process, but it is a

---

12 What an unrelatable level of self-confidence.

relatively straightforward and well-documented one, and we'll talk more about it in Section 3, "Scaling Teams".

High performers struggling with change are much more challenging for managers to handle: one, because you may feel like you need them more than they need to stay, which makes it scarier to push back; and two, because high performers tend to have a lot more social capital in the team, and if they are skeptical, others will be too.

So the high performer everyone on your team loves and learns from, who you worry the team couldn't function without, who turns up to your 1:1 and tells you everything that is going wrong? This is the person you need to get buy-in from. It will be hard, but getting them onside will really help you be effective, so take the time. Hone your explanations on them, hear them out, and work to earn their trust. Show them that you will bring them (and others!) with you. Just accept that for this person during this process, it may feel like everything you do is too little and too late. However, eventually, when you've succeeded together, their feedback and validation will be the thing that tells you that you've finally made it out the other side.

ALL PROCESS IS CONTEXTUAL

Sometimes people look at teams and diagnose problems as an absence of process rather than an absence of values, cohesion, or delivery. They enact some kind of process—a weekly meeting or a review process, for example—and then wonder why people aren't buying into it and why it wasn't the miracle they hoped.

Lack of process is rarely the actual problem but rather a symptom of an underlying problem, which is what you need to address. You may still create the same process, but you should ensure that the process is a contextual answer to a contextual problem in service of an outcome, and not just an added process for performative reasons. We'll talk more about building effective, outcome-driven processes in Section 4, "Self-Improving Teams".

When I joined the mobile team at Automattic, there were no standups. But this wasn't the problem. The problem was that some people didn't talk to one another, surface problems, and offer or ask for help. Others did, and so for those people, a standup seemed like a worthless addition of process. But getting everyone to a place where they would talk to one another, surface problems, and offer and ask for help was really important.

Retrospectives are another process that change managers will be tempted to impose on teams that don't already do them. But while retrospectives are a worthy goal, they are rarely the place to start. When teams struggle to do retrospectives, often the problem isn't the lack of retrospective but the lack of psychological safety that a productive retrospective requires. That may be due to lack of trust or cohesion on the team or to other problems that have not been addressed. You need to address those things first; otherwise, the retrospectives will not be productive, and the process will be for naught.

Standups and retrospectives are great examples because while you (might) be able to force people to show up and say something, you can't force meaningful participation. It may feel like you spend a ridiculous amount of time getting people onboard with something that is a very standard practice, but sometimes that is the point. If you expend that energy on one process, and people see the improvement that results, they will be more open to—and hopefully start suggesting improvements themselves to help with—what comes next.

YOU WILL MAKE MISTAKES

The biggest change management mistake I made was in 2017, when I reorg'ed my team twice in around a four-month period. First, we aligned ourselves by time zone rather than by project. We made that choice later than we should have done, stalling out of fear and things we hadn't fully dealt with.[13] We expected that structure to work for some time, but then we hired more people than we had planned. We also hired more individuals who were located in a different time zone than we had historically, which broke the structure sooner than anticipated. So we had to enact a second reorg to fix the team structure in a way that incorporated the extra people and the additional time-zone constraints. I felt terrible about this, knowing firsthand how disruptive reorgs had been for me as an IC, and when I communicated what was happening to the team, I was honest about why it had happened and the gap between what we'd expected and what actually happened, and I apologized. People's reactions surprised me: instead of criticism and complaints, I got empathy and understanding, and people shifted into the new structure peacefully, without a noticeable dent in output. I worried that admitting this failure would undermine me, but the mistake was so obvious that being anything other than direct about it would have only made it look worse. So I was direct, and the people I worked with surprised me in the best possible way. Whenever I am hesitant to admit that something didn't go to plan, I think about that, and it pushes me toward honesty and transparency as a way to build trust with the people I work with.

---

13 Yes, that was the "too much empathy" communication failure.

There are certain practices and principles of leadership, but it is an art rather than a science, and this is even more true in the work of team transformation. If we're not sociopaths (and I do hope none of you are sociopaths), we allow people on our teams space for their mistakes, and part of that is acknowledging that despite our best intentions, we will make mistakes too. We need to admit them honestly and openly and apologize for disruptions we have caused. When we admit our own mistakes, we build trust. When we don't, it demolishes trust. To not acknowledge our mistakes is a form of gaslighting that undermines everything else we might accomplish.

There's no way around the mistakes—only through. So learn as much as you can from them and admit them to as many people as you can. You'll do better next time, and your team will be that much more ready to believe that, too.

Story

For most of my career, I have been an individual contributor who was mostly focused on deep technical work. As a result of the nature of the projects I worked on, I never had to effect any kind of change in people or in systems. About five years ago, I started managing people, and effecting change became part of my job overnight.

It started in small ways: showing my reports the potential behind a new audacious project that looks a bit boring in the beginning and convincing them to work on it, coaching them to develop new skills, and so on. At first, even small changes didn't seem plausible to me—what would I say that's not already obvious to people? But it turns out that people love to be seen and then challenged and guided, and they often are missing information that shows the big picture, and so I was able to effect good changes.

I started finding transformation as a concept really fascinating and started advocating for larger changes around me. To name a specific instance, there was a particular deficiency in our systems that impeded the velocity of launching new features yet was silently accepted over time. At first my advocacy for change did not yield any results—I was a lone voice and too new. As I grew credibility by delivering work, I started getting heard but not enough to sanction the changes I wanted. It was a particularly draining period that I found difficult to persevere through. I took frequent breaks and revived my advocacy in new forms and kept going until I found some allies. From that point, everything got better. We joined forces and gathered more data. Our combined voice and the data demonstrated that the problems were worth solving and worthy of a long-term strategic investment in the infrastructure. We took ownership of the solutions and intentionally drove change through the org, speaking to leaders across to get their buy-in. In addition to the technical change, we also advocated for a cultural change to take ownership of feature velocity.

This was a big learning moment in my career in that I realized that I can drive really big changes. I recognized some strengths in myself, such as the ability to be firm and resilient, to have clarity in solutions, to apply systems thinking to tease apart the contributing factors to problems. I also noticed some failure modes: getting discouraged when I don't succeed in influencing and finding it arduous to repeat myself. Once I developed some tactics to overcome the failure modes, I realized that change is often possible, even when it looks nearly impossible.

Some important lessons I learned:

- Never assume things are obvious to everyone but you—they are not. Speak up and add your perspective.
- Partner with allies—you lift each other up.
- When you feel like giving up, take a break and try again later, maybe in a slightly different way.
- Gather data and quantify the impact of the change you want to see.

—*Nandana Dutt, engineering director, Google*

Your Action Plan for Self-Management

We've covered a lot in this section, so here's a suggested path for implementing or improving your self-management. This can require many changes that will take time, so make sure you give yourself room to work through it. This list is long, and maybe a little overwhelming, but you're not supposed to go through it all at once—review your next action and give yourself some time to take it, and then come back to it again later.

Step 1: Assess your energy management and make your plan.

Maybe your energy management is fine—if that's the case, I'm happy for you. But hopefully, you have identified some areas that could use a little attention. Where to begin and how to prioritize?

Step 1.1: Check yourself for signs of burnout.

Are things OK but could be better? Or are you heading toward a bad place? This will help determine the scope of changes you need to make.

1. If you're heading toward a bad place, some dramatic shift is needed, and heading things off before they degenerate is a priority.

    a. Reach out to people who can help—a close friend or a coach—and set up time together.

    b. Work through the list of factors and identify the biggest issues.

    c. Make a short-term mitigation plan that will prevent things from getting worse.

    d. Identify and move toward the larger changes that will take more time.

2. If things are OK but could be better, it's a great time to start making some minor interventions.

    a. Identify the top thing you would like to improve or change your mindset on.

    b. Identify and set up some structure that will help you do that.

    c. Check in with yourself in a month or so: are things any better?

    If yes, move onto the next thing.

    If not, root-cause why the change didn't work and try again.

    If multiple efforts haven't been effective, it is probably time to seek out some additional support or bigger change.

3. Everything's fine! Fantastic! Go to step 2.

Step 1.2: Identify where you're being useful rather than strategic.

Running around trying to be useful can fill your time but undermine your effectiveness and ability to think strategically. Escaping the usefulness trap is key to making bigger changes.

1. Identify if you've been doing anything that has undermined your effectiveness.

    a. These are the things that are most critical to change.

2. What activities can be immediately dropped? Drop them!

3. What activities need to be transitioned to others?

    a. Figure out who and make a plan to delegate.

Step 1.3: Identify activities that are disproportionately draining.

These are the things that are most corrosive to your well-being. Probably some of them came up in step 1.1 or 1.2, but there may be more.

1. Do an energy audit as outlined in "Managing Energy Versus Managing Time" on page 84.

2. Make a plan for the following week that considers energy rather than time.

3. Identify longer-term changes to work toward, and factor those into your to-do list.

    a. If you do one thing each week for future you's effectiveness, what will you start with?

Step 1.4: Determine the core components of your self-care routine.

It's so easy to get caught up in what other people expect of us and put ourselves last. What do you need to operate at your best?

1. Make small changes (like blocking lunch on your calendar).

2. Build support structures and habits. Get creative!

    a. For example, join a reading group to ensure you read books regularly.

    b. If exercise is core for you, maybe you need to hire a personal trainer in order to exercise consistently or set some shared goals with a friend.

Step 2: Identify opportunities to improve.

We can all get better, but sometimes it's hard to know where to start or how to find time. Match your areas of improvement with things your team needs to function better.

Step 2.1: Identify gaps with your manager and how you approach managing up.
It's easy to let your manager determine your relationship and accept (or get frustrated by) whatever you get from them. You have more power than you think!

1. Review the components of management: what are you getting, and what are you missing?

2. Identify what you can get elsewhere.

3. Focus on what you need from your manager.

   a. Try and refine some of the useful questions.

   b. Rebuild your 1:1 agenda to help you get what you need.

Step 2.2: Mine for implicit feedback.
When you're not getting explicit feedback, or the explicit feedback you are getting is not actionable, you can always turn to implicit feedback.

1. Consider the growth framework and identify what you can improve or what has recently changed.

2. Think through how your job most recently evolved.

   a. What can you learn from that?

   b. What went well?

   c. What did you struggle with?

3. Review the components of management and identify gaps in what you're giving your directs.

   a. What do you need to adjust or be more explicit about?

   b. Consider updating your 1:1s to prioritize topics that have been getting missed or to ensure that all components get covered.

4. What are your failure modes, and how can you improve?

   a. Do a values exercise and consider how they showed up in recent mistakes.

   b. Take the StrengthsFinder test (*https://oreil.ly/VVvhv*). Do your strengths match what you are doing? Are there flip sides of those strengths making it harder to have the impact you want?

c. Take the saboteur quiz (*https://oreil.ly/gWkiP*) and start paying attention to when your saboteurs show up.

Step 3: Define your strategy.

Part of being effective is knowing where we are and where we are going. Once you've sorted through the noise (overwhelm) and figured out what's important (direction), it's time to make a plan.

Step 3.1: Determine your team's current state.

Figuring out where we are today is core to making a plan. Consider where your team is at versus organizational needs. What are the proximate objectives?

1. What is your team's current state, in terms of people/scope?

2. What are the current bottlenecks?

    a. What needs to change immediately?

    b. What will need to change for the team to meet the organizational need?

Step 3.2: Match the style to the situation.

When looking at current bottlenecks, you may see opportunities to try different leadership styles or communication strategies. Is the default democratic style resulting in decisions not being made? Is the team worn down by pacesetting?

1. Assess your leadership range.

    a. What are your natural go-tos?

    b. Where are you holding yourself back?

2. Identify where different leadership styles may be more effective.

    a. What support structure or help do you need?

3. Consider the communication failures: can changing communication style or frequency help?

Step 3.3: What challenges are coming?

Are significant changes in order? In Section 3, "Scaling Teams", we'll talk about scaling teams, and in Section 4, "Self-Improving Teams", we'll talk about getting teams delivering better.

Section 2 Summary

Congratulations! You are now the manager of yourself.

Wait...this was always true. But hopefully, whatever the org chart says, and however great your manager is (or isn't), you have some new tools to help you be effective and evolve your style and role.

In Chapter 5, "Energy Management", we talked about shifting your mindset to focus on managing *energy* rather than *time*, avoiding energy-draining activities (and running around after everyone!), and navigating and extracting yourself and others from a place of burnout.

In Chapter 6, "Defining and Adapting Your Role", we talked about what your role is, what you need to get from your manager (or elsewhere) to be effective, and how to start getting strategic when it comes to managing a team. We also talked about implicit feedback—what it is and where to find it—and how to approach evolving your role over time.

In Chapter 7, "Expanding Your Leadership Range", we talked about defaults and strengths, controlling weaknesses, and how to improve our effectiveness by identifying our strengths, controlling or mitigating our weaknesses, and paying less attention to our saboteurs. We covered the different leadership styles, when they work best, and how to find opportunities to try them out.

This is the end of the first half of the book, the part where we focus on *you*. Next, we will move on to the teams you support and how you make them more functional and more effective. In this section, we talked about identifying what you need to do to meet your own needs, the needs of the team, and the needs of the organization. Now we'll get tactical on how to actually do that.

PART II

Team

Engineering teams have three primary components: people, projects, and process. Strong teams include people who are individually doing well, and they deliver business value, predictably and effectively.

I don't include process in that statement deliberately. Good teams are not defined by their processes; good teams use and evolve process effectively to meet the organizational need.

In the first half of the book, we focused on *you*. Your career, your role, your effectiveness. Now it's time to move onto the team you are responsible for or part of.

Teams exist to serve a purpose. Engineering teams exist to deliver business value. That business value is directly tied to why engineers are so well compensated.[1] Whether it's a team focused on developer experience (DevEx) and internal users, or a consumer-facing team, that team produces value—whether it's measured in acceleration of other teams (hardest to quantify) or sales (easiest to quantify). When we grow teams (people)—as we will discuss in this section—we do so in service of the business goals. In "Self-Improving Teams", we will cover how to most effectively serve those business goals (projects) through building a team that is *self-improving*. Process is sprinkled throughout as relevant and useful.

In all of this, we will be focusing on how teams *scale* while being *effective*. This is deliberate because scaling is often what causes processes and effectiveness to break down. As such, part of scaling effectively is adjusting as we go to minimize "growing pains," or, as we are now calling them, "strategic

---

1 It's wild to me how little this is talked about. Engineers—understandably—like being highly paid, but that is only possible when the business is profitable enough to support it. The revenue-per-employee model is common, and those numbers matter in terms of business sustainability and the stock market. You don't have to like or agree with capitalism to understand the levers and how they can affect you and your team.

deficiencies." These issues do not show up exclusively during team scaling, but they are often most noticeable then. You can improve effectiveness even if a team is not scaling, and if a team is shrinking, you will have to.

This second half of the book is meant to help you as a manager to build stronger and more effective teams—to help you develop and execute on the strategic elements of the job that we started talking about in Chapter 6.

If you are not a manager, this book can help you give language and form to the problems you notice but don't have the power to fix directly. It should help you take ambiguous feelings like "I don't think our hiring process is working that well" or "Delivery seems worse than it used to be" and make them more concrete, giving you ideas of what would help those things work better. You may have more influence than you think in terms of effecting change, and a better understanding of what good looks like will support you by giving more useful, constructive feedback to people who are ultimately responsible or have more power to effect that change. My warning here is that this kind of activity can result in you being pushed onto the management track, but my hope is that one day we will be able to want and expect teams to be better without having to run them ourselves.

Scaling Teams

In previous years, we experienced boom times of zero interest rates and unicorn (greater than $1 billion) valuations based on user numbers rather than revenue.[1] Back then, the number of people employed at a company was a vanity metric; growth numbers mattered more than business fundamentals, and organizations often used to "scale teams" by adding more and more people to them. Then, much later, they'd only intervene when things started to break down. Managers would be hired, objectives and key results (OKRs) would be introduced, and leveling systems would be rolled out. There would be attrition, but eventually some semblance of order would usually be reached—at least until the market shifted, questions about productivity ensued, and a layoff was announced.

This approach to scaling was always pretty wasteful. Adding a lot of people very quickly usually meant that hiring was not well thought through, meaning there would need to be more performance management (HR would typically also be unprepared for this, either because of hiring in late or focusing on culture), and onboarding would be chaotic and inconsistent, making it difficult to know where to attribute problems or how to address them. If diversity, equity, and inclusion was a priority, the DEI numbers might improve, but if it wasn't, a small homogenous team would grow to be a large homogeneous team, and that problem would become even more intractable.

People management is a key part of the strategy for scaling a team. People management is how people are hired, how they are onboarded, how they are supported, how they are developed, and how they are rewarded. While direct people management is a piece of that, overall the point is creating an *ecosystem*

---

1 One example is Fast.com (*https://oreil.ly/4JFTJ*), which raised $124.5 million and spent it all in about 2.5 years. Another is Fab (*https://oreil.ly/yeP6Q*), which went from the "fastest growing startup in the world" in 2011 to reach a $500 million valuation in 2012...and was gone by the end of 2014.

that supports strong people management (in all these areas), which ultimately results in faster development (e.g., reduced time to onboard).

For instance, if we think about developing managers, the ecosystem the manager exists within is at least as impactful as the relationship between the new manager and their manager. People who are well integrated into the ecosystem are more invested in supporting it. They also have:

- Clearer models of what "good" looks like, from multiple different people
- Stronger support networks
- Shorter feedback loops (i.e., someone nearby is more likely to notice something amiss and more likely to tell them)

At a general level, you can see how this applies to other things under the people management category, such as hiring and onboarding:

- People who feel more secure in the ecosystem are more comfortable hiring people who will add something extra to what they do (whether that's diversity or just a different skill set).
- People who have a stronger model of what good looks like will feel more comfortable evaluating others on defined capabilities.
- People who have a stronger support network will feel more comfortable with the idea that there is capacity to support people in *trainable* skills.
- People who feel secure are more comfortable questioning whether things are working.

Having done multiple team transformations, I've repeatedly observed that one of the most impactful things that emerges from a team transformation is a different ecosystem. Sometimes, the processes on a team are very similar and the projects are basically the same, but the entire vibe on the team is different. A strong predictor of retention (both regretted and unregretted) is how integrated people are into the new ecosystem. How well new hires are integrated is also a predictor of retention, and this is a piece of why onboarding is so important.

Like any ecosystem, the healthier a team is, the more diversity it supports. The lazy shortcut is cookie-cutter hiring and homogeneous teams—I don't believe that works in the long term, but what do I know? I've never done it. As such, you'll see a strong theme of diversity and inclusion throughout this section. Some of it may be challenging to you, and if that's the case, I invite you to sit with those emotions and see what learning is there for you.

Also note, there will be many things missing here with regard to diversity and inclusion. My perspective and experience is that of a White woman from a wealthy country with a computer science degree: highly privileged in terms of what it means to be a "marginalized person in tech." There is much more out there, in particular from women of color. If you are early in your journey toward better understanding diversity and inclusion, I'd encourage you to keep learning well beyond what is covered here. Project Include (*https://projectinclude.org*) is a great place to start.

Fundamentally, this section is about how to set up teams to grow. We'll talk about how to build effective hiring processes that scale, what levers you have to improve team diversity and how to use them, how to drive effective onboarding, and how to identify, grow, and reward the people and behaviors that align with a well-functioning team.

Scaling is both a people problem and a process problem. To scale, you need to add people, build out functioning leadership structures, and keep people aligned with the organizational values, which will themselves need to evolve or be clarified. The way that small companies function can cause early employees to mistake constraints for values. For instance, a lack of people management in a small organization may be a function of lack of need (and money), but this is not the same thing as an organizational belief that managers are "useless overhead."

Process is similar. Lightweight or nonexistent processes can be adequate in small companies and situations where you have a couple of people who are very aligned handling things together. Once scale moves beyond that, it becomes necessary to make processes clearer, more replicable, and less dependent on core people. This is why many incoming leaders focus on evolving or implementing process—with mixed results. Good process does not exist in a vacuum; good process—like strategy—is contextual and works from the realities of the situation as is while effecting change. Part of this is providing a good *why* for changes to be implemented. As a result, I have tried to outline in this section the mindset and priorities for effective processes for scaling, not the specific processes themselves. You will need to adapt these suggestions to the environment you are in, incrementally, to use them effectively.

8

Hiring That Scales

When given the opportunity to establish a hiring process, we're all biased to advocate for a process in which we would be successful ourselves. In hiring, this plays out in two main ways: "A" players build monocultures, hiring people just like them, and "B" players hire "C" players, hiring people who won't threaten them.

In other words, top performers too narrowly define what top performance is, and OK performers hire mediocre people; in both cases, they're making selections that bolster their own position.

From the candidate perspective, the hiring process is the start of someone's experience with your team. In an ideal world, this connects all the way through their employment with the company: the competencies you evaluate are consistent with your promotion process, the ways in which you evaluate them are consistent with the way you work, and how you approach that evaluation demonstrates the kind of team that you are.

Hiring is an eternal struggle, especially in software, because of (1) market changes, (2) the internet being full of bad advice, and (3) the impossibility of complete data. Regrettably (thankfully?), we cannot A/B test humans, and we will never know how the people we didn't hire would have done if they had been hired. Even the data from the people we *did* hire can be noisy, with no clean control based on their onboarding support, manager, first projects, and personal lives.

In this, probably the most tactical chapter in this book, we'll cover how to build an engineering hiring process that scales while also improving team diversity, including how to set up your metrics so that you can prioritize interventions effectively.

Assessing the Current State of Your Hiring Process

To know what to adjust, the first thing you must do is assess how the process is performing. Let's divide that process into two categories: useful metrics and useful feelings.

METRICS

The core metrics that will help you assess the health of your hiring process are diversity metrics and process metrics.

Diversity metrics

There are two big challenges in measuring diversity. The first is that it's hard to do: diversity is about including more of the full spectrum of humanity. How do you put hard numbers on that? Especially if you're hiring only a few people each year? What's reasonable to expect? What even *can* you measure without being weird and creepy?

The second is that you manage what you measure, and this can result in an unhelpful direction. "Diversity" becomes "women" (because it's measurable and possible), which then becomes "hire junior women" because that is the easiest way to meet that goal. However, there's more to it than that.[1] Racial diversity is truly important. We're not building meaningfully more diverse teams if we add a couple more White women to a sea of White men. However, racial diversity can be very difficult to measure at a smaller scale, even more so than gender diversity.

> ### Warning
>
> In the United States, it is very difficult to collect meaningful diversity data about your pipeline without introducing legal risk, and you cannot use demographic information in decision making. Other countries have different constraints; you will need to see what information you can legally obtain or request.

---

1 IBM has one of the longest-running commitments to diversity. Starting on September 21, 1953, when Thomas J. Watson Jr. (president of IBM at that time) wrote in Policy Memo #4, "It is the policy of this organization to hire people who have the personality, talent and background necessary to fill a given job, regardless of race, color or creed." This was nice as an idea but limited in efficacy, especially in leadership roles. As a result, in 1993, when Lou Gerstner took over, he implemented a comprehensive DEI strategy (*https://oreil.ly/hzCJw*), aiming to change not just entry-level hires but also the overall demographics of leadership. By 2005, the number of female executives worldwide had increased by 370%, but it wasn't until 2012 that Ginni Rometty became the first female CEO of IBM. The overall lesson we could take from IBM is that there is no "success," just ongoing effort and iteration to find what works.

Having acknowledged those challenges, we still need some metrics to ascertain whether what we're doing is working. Attempting to measure diversity is extremely difficult: diversity refers to more than gender, and gender is not binary. However, people who identify as women make up around half of the population (making them statistically significant and measurable), and so I advocate for using gender as the best available proxy metric around how your hiring process is working. This is most useful when you set thresholds for which being *below* a metric is indicative of a problem.

In engineering, I follow the heuristic that a process that hires less than 20% women should be considered to be fundamentally broken. And if it's below 30%, it needs work. Note, this is *a process*. The focus needs to be present at all levels. Even if you hire 100% women junior developers, if your senior developer process hires only 15% women, you still have a process that is broken that needs to be addressed.

The good thing about proxy metrics is that they don't have to be completely accurate and are mainly used in aggregate. Ideally, I like to be able to generate a report that can show me gender breakdown at each stage in the process, but whether you can get that will depend on the constraints (location, size) under which you operate. You can also get the information you need by using a script and an API (such as Genderize.io (*https://genderize.io*)) to make reasonable anonymized estimates of gender breakdown at each stage. If there's a stage where the conversion is not comparable, then that's a good stage to look into to discover how it can be improved.

Process metrics

Diversity is a metric that indicates quality of overall process outcomes, but we can still measure conversion of different stages. Combining those metrics with diversity in stage can suggest where bias is most prevalent.

Withdrawal rates and reasons

Withdrawal rates and reasons tell you a lot about whether people experience your hiring process as a good use of their time, and if not, why not. It's worth considering the reasons and what you can do about them:

Compensation

It's wasteful to have people withdraw once in process because of compensation, and this is why it's a good idea to put your compensation ranges on your website (this is a legal requirement in some locales). However, if this is a thing that is happening, it's worth tracking and reporting on it in order to advocate for change. If your offers are under market, your current employees are likely being compensated under market, and that can trigger attrition.

Time commitment

It's understandable that you want to be confident in your hires, but if you are consistently getting the feedback that your process is too arduous a time commitment, it's worth looking at how you can reduce it. We'll talk more about this in "Improving Evaluations" on page 174.

Experience

Most people won't say "bad interview," but if too many people withdraw after a particular interview, worry about the impression those interviews are giving people. Glassdoor reviews are another way to learn this. While, yes, many of those reviews are angry and bitter, there are sometimes enough comments around a particular theme that you can extract a valid pattern.

Job role

If you get feedback from people that the role they find in interviews does not match the job posting, it's likely that the job posting needs to be revisited for accuracy.

Time metrics

Time metrics tell you how long the process takes and where that time is spent.

Time in process

This is the amount of time it takes someone to get through the hiring process start to end. When people interview because of the specific job (such as sourced candidates who aren't actively in other processes or mission-aligned, highly motivated candidates who visit your job page regularly), this may not matter as much. When hiring at low volume, selecting for this profile may work well enough for your needs. However, once volume increases, this metric becomes increasingly important. When people are actively looking for a new job, they'll typically start several processes at once. The longer yours is, the more likely it is that they will get a compelling offer elsewhere and drop out before your process is complete. You may think you're offering something truly special and unique,[2] but even if you are, you can't realistically expect people to turn down a concrete job offer they're happy with for a "maybe" with your company. There's as much need for your time requirements to be competitive as your compensation.

Time to fill

How long does it take from deciding you want to hire for a role to filling it? This is where you can see the bottlenecks. Do you spend two months waiting on approval to fill the role? Does it take two weeks to get the post on the website? Does it take six months of interviewing to find someone? The overall length is useful information, but more useful is identifying specific process bottlenecks within it. It may be possible to dramatically cut overall time to hire while focusing on only one specific aspect of the funnel—a huge win.

Confidence

This is how strongly you feel about people at each stage in the hiring process. I really like to divide candidates into the buckets of strong yes, weak yes, weak no, and heck no, provided these come with a description as to why.[3] Often, this overall impression reveals the uncertainty about whether that individual would be successful.

For example, an interview might be a "weak yes" because the candidate didn't go deep enough technically for the interviewer to have confidence they will pass the code test. Rather than bring in an expectation of an assessment that's done elsewhere in the process, it makes sense to move that person forward (and then calibrate on the result for a more meaningful signal). However, if it's a "weak yes" because it's unclear the person has the experience required to be successful, it can be worth reviewing the other information you have and not wasting any more of anyone's time if it's very likely to be a no. So if you're interviewing someone for a director-level role and discover the "VPE" title they have really just means tech lead of three people, that's probably enough information to pass.

---

2 And how are you convincing people of that? Another difficult problem!

3 "Heck no" needs to be used judiciously. "They didn't use my preferred syntax" is not a valid reason. "They made racist/misogynistic comments" is.

Conversion metrics

These metrics indicate success and help you understand how strong a filter every stage is and whether that is proportionate. As a general rule, conversion should increase as you go through the process. Open job postings typically generate a homogenous pool of unqualified applicants (which is why sourcing is critical for improving diversity at the top of the funnel), and you should expect to progress relatively few forward. The more time you spend on someone—and the more time you ask them to spend in your process—the more confident you should be in their eventual success.

Pass rate

This metric shows how likely someone is to progress to the next part of the process. Low pass rates in one place can indicate problems elsewhere.

A Process Gone Wrong

At Automattic, we had a process that began with resume review, interview, and then code test. We tracked our metrics, and after a period of things improving, we saw a drop in success rate at code test. However, nothing had changed in the code test, and we were pretty confident it worked, so we started working backward. We looked at interviews: were interviewers being too lenient? The pass rate there was about what we expected, so it hadn't indicated a problem, but when we looked into it, we found that interviewers were complaining about having a lot of bad interviews with people who seemed clearly unqualified. It turned out that the recruiters were not well calibrated, and they were advancing too many people to interview. Then, the problem continued at the interview stage, where the volume of bad interviews made the mediocre interviews look good, so those people were advanced and subsequently failed at the code-test stage. Even worse, the interviewing capacity was more than fully utilized, and people were having to wait a long time for an interview, which resulted in more no-shows and longer time in process. So many problems—but it all started at the *beginning* of the process, with the resume-review process.

To address this, we enforced stricter resume review and reduced immediate interviewing capacity (so people couldn't book further than one to two weeks out). By *lowering* our pass rate at resume review, both pass rates later in the process and overall efficiency improved. By setting a firmer boundary on capacity, we kept things moving better. It's fine, and even a good idea, for recruiters to have the occasional wildcard for people they feel strongly about, but this certainly shouldn't be applied to the majority of people in your process—and those wildcards should only be allocated *if there is capacity to interview them*.

Accept rate

A high acceptance rate can seem like a good thing, but in reality, that means you have a process that excludes everyone but the most committed. According to Lightspeed's 2018 Startup Hiring Trends report (*https://oreil.ly/Jlcsp*) (see slide 26), a typical acceptance rate for engineering is between 65% and 69%.

A low acceptance rate suggests you're not making a compelling offer, whether in terms of compensation or the job itself.

If you find a low acceptance rate, the first thing to check is always compensation; market rates evolve, and if your package is not compelling, people will turn it down. It's worth asking why you are finding out that you're not meeting expectations at the end of the process. Have these conversations up front; it's more efficient. It's important to be aware of your market and what competitive compensation looks like. For example, working for distributed-first companies, I don't expect to hire in the most expensive markets like the Bay Area or New York City (and I don't recruit in those markets). If someone turns down based on comp in an expensive market, I'm not too worried. If someone turns down based on comp in a market I expect to be able to recruit from, I pay attention.

Use the data to negotiate more realistic salary bands across the board, ensuring your salaries are in line with the markets you expect to hire, rather than trying to create exceptions per person. It's understandable that you feel short-term pressure to hire, but it does not help you in the long term (or even the medium term) to have one person who is compensated significantly more than others on the team. Equally, it would be a gross waste of time to wait until you have multiple offers turned down based on comp. By having the compensation conversation earlier, you can get the data earlier and use other sources of data to make the case that you're not matching the market. This will also make a case for correcting compensation packages internally; otherwise, you will have even more hiring to do. It may take time, or it may not be possible to change the compensation model, in which case you will need to look at the levers you do have.

The next thing to consider is what impression of the job and team the hiring process has given. It's worth asking for feedback. These folks have spent a significant amount of time in your process, and if they turn down the offer, you can ask for a call to better understand their rationale and get some feedback. You can leave the relationship on warm terms.

Finally, play a long game. In this industry, there's a good chance you'll be hiring (at least somewhat) forever. If you were confident about hiring someone and are still hiring, why not be willing to reopen an offer later? I think it makes sense to let someone who has gone through the interviewing process pick up where they left off and reopen an offer within approximately one year of interviewing (barring significant change to process and some consideration of what they have done in the intervening period). Sometimes people get to the end of a process and instead decide to stay where they are, but they may be ready to move later, so keep the door open. I've also found value in maintaining contact with the person to reiterate that we'd still love to bring them onto the team. It might not be the original role they interviewed for, but that's OK.

There are a few key questions you should ask when closing the hiring process with a candidate:

What do you want from your next job?
Be honest about the ways in which this job can—and can't—give them that. It's best all around if someone makes the right decision *for them.*

Do you have any questions for me?
A good hiring process should be two-way. Allow the person to get the best sense possible about what the job will actually be like. I like to tell people that I believe everyone should get the opportunity to interview a potential manager and that I'm giving them that opportunity to do that with me. If I won't be their manager, I tend to frame it around "I want you to ask me anything you need to know in order to feel *great* about your decision."

Is there anything you want to know about diversity and inclusion?
Sometimes people don't like to ask about these topics without invitation and limit their impressions to backchannels and what they can find out on the website. I've found that explicitly putting this topic on the table is really helpful for surfacing people's concerns in this area—or at least being more confident that wasn't why they said no.

Success rate

This is how often the people you hire are successful; at its simplest, this is whether they meet the expectations of the role and team you hired them into. It may seem that 100% is the goal, but that indicates that you're ruling out people who *could* be successful because you require too high a confidence level. If your hiring is resulting in a homogenous team, that is all the more reason to dig into this.

If people are not being successful, it's worth digging into what you can learn from that. I spent a lot of time revising a leadership hiring process after discovering that it was a poor indicator of success, in part because it was too focused on someone's individual work rather than how they were able to lead teams to deliver effectively. We will talk more about some additional considerations for leadership hiring in Chapter 10.

FEELINGS

Do people on the team *think* the hiring process is working? And if not, why not? Do they want to participate in it?

Often when people feel that a process is unfair or unpredictable, they feel anxious that they themselves would not make it through the process. This anxiety is unhelpful, in the way that any workplace anxiety is wasted energy, and specifically also because it usually makes them either more lenient or more harsh.

A first step toward helping people feel more comfortable with hiring decisions is standardization (more on this in "Improving Evaluations" on page 174). When people understand what is being evaluated and why, they are much more comfortable being part of the process and making a recommendation.

When I took over the mobile team at Automattic, the hiring process was arduous and arbitrary. As part of the interviewing process, they would send candidates off by themselves to perform really challenging projects—things people already on the team wouldn't even think about approaching by themselves, such as figuring out solutions for "offline support" or "multiple logins." In other words, it was not realistic to succeed with these projects. As a result, the measurement for success wasn't completing the project but rather was a candidate's willingness to suffer through an impossible task. Thus, anyone who managed to get through this process later went on to inflict the same type of bad hiring process on future candidates.

I was appalled by this mindset and did a number of things to change it to one that was more constructive and fair to candidates—and representative of a more collaborative way of working. One important aspect was to draw attention to this way of treating people. I pointed out, "It's a hiring process, not a hazing process" and asked people what they thought would set the candidate up for success or be a reasonable thing to evaluate. Most people quickly adapted to this kinder mindset, and they were more comfortable participating as a result. As it turns out, most people do not want to watch someone else fail on a project they weren't confident *they* could do alone.

The Myth of the Hiring Bar

It is not possible to change a hiring process without someone putting forth a fear that this means "lowering the bar."

There is no hiring bar. The entire concept is nonsense. How can you distill a human's capability, potential contribution, and weaknesses into a single measurement that means "above" or "below" a hypothetical bar? We use someone's height to say whether someone *can* ride a roller coaster, but that number says nothing about whether they want to, whether they would enjoy it, whether they ride it just once or again and again and again. In short, this fictional bar doesn't measure a candidate's potential.

There is what determines success on your team, and then there is what you decide to evaluate. Evaluation must be consciously designed and mindfully applied, or else people pattern match rather than actually evaluate competencies. It's easy to fall into the trap of expecting less from some (and thus giving them more slack) while applying more stringent evaluations to others. This demonstrates why a lack of diversity is a sign that a hiring process is failing: it is a sign that evaluation is more based on people thinking, "I know it when I see it" than clearly articulated competencies with fair evaluations.

The bar that is clearest in hiring is the effort bar: the effort required to undertake the process at all and the effort to be successful. This is something we can distill down to some kind of range and determine whether it is reasonable—or not. *The higher the effort bar, the more people are discouraged or excluded from participating.*

You can try and make a compelling value proposition to overcome that discouragement, but if people don't have the time to participate because of other commitments in their lives or existing jobs, then they simply can't participate. It's wild to me to have an overly difficult process that favors people either being between jobs or who are willing to put minimal effort in on their current job while looking for the next one. You rule out people who are casually looking (most of the engineering job market) or who take their commitments to their current team and employer seriously.

Often when people talk about "raising the bar," they actually mean raising the effort bar—and selecting for those willing to put that effort in. It's worth making that explicit and thinking through the potential consequences. *Asking candidates for more work doesn't mean we get more qualified or engaged people but often the opposite—we just get people with more time.*

Prework for Improving the Diversity of Your Team

We talk about "diversity and inclusion," but perhaps it should be "inclusion and diversity" because **inclusion needs to come first**. Don't hire people into an environment where they can't be successful. On a practical level, it's a waste of everyone's time. On a human level, it's harmful.

Inclusion is about how welcome people will feel if they are part of your team. Consider if you have to be a certain "type" of person (outgoing? willing to participate in a team social culture that revolves around alcohol? hypercompetitive?) to be successful on the team. Are these characteristics really necessary? Would your team benefit from people who are quieter and more thoughtful, more collaborative? Would those people feel welcome—and be able to be successful—if you hired them?

A good rule for inclusion prework to diversity is to stop doing things you would have to change if the demographics of your team better reflected the demographics of the world. If you find yourself watching interactions or jokes and thinking they wouldn't be OK if there were women/people of color/LGBTQIA people on the team...you need to shut that stuff down *now* if you ever want to have women/people of color/LGBTQIA people on the team. Or, just because it's the right thing to do.[4]

---

4 Not only women are offended by sexist jokes, and not only people of color are offended by racist remarks, so it's a mistake to think that those kinds of comments are OK when the people who are targeted aren't present. They aren't.

For an easier example around inclusion, let's take time zones. If you want to hire more globally, it's worth moving to more asynchronous communication *before* that happens. It's not a great experience for any new hire to feel like they forced the team to adjust to a different set of working hours. That adjustment also takes time—time in which it's harder to set the new hire up for success.

As another example, if you realize your team is hypercompetitive, for example, you may want to think about how that is encouraged by the hiring process and the team environment. How much of that do you have control over? If it's created by your promotion process, can you influence it?

Really, inclusion is just considering who should be able to be successful on your team and making choices to maximize the chance that they can be. The worst way to approach inclusion is to hire an "only" (the "only" woman on the team, the "only" person of color, and so on) and fix their struggles one by one while making them feel isolated and blamed. It's probably better not to try at all.

Once you believe your team is one where under-indexed people can be successful (and happy!), you can start building your pipeline.

BRAND AWARENESS

Some companies engage in what I term "diversity as performance art." This is a PR exercise of "diversity" work in order to appear better rather than to be better. It can help diversify the pipeline but typically only works to a certain extent, most often at junior levels. The focus isn't on structural changes to improve equity or enabling under-indexed people to be successful, and that's problematic; it can make DEI work another form of office housework—somehow an expectation but not one that advances the career of anyone doing it. It is more focused on the appearance of progress than actual progress, making it inherently unsustainable in driving meaningful change.

For a more meaningful and sustained impact, look for ways to give under-indexed folk more recognition for their actual work—what you actually pay them for—being an awesome engineer, or product manager, or designer, or whatever it is they do. Send them to industry events rather than just affinity events, and showcase their work on your company blog, not just your "diversity and inclusion" page.

Having happy and successful under-indexed people on your team is the best case you can make to other under-indexed folk that they too would like to be on your team.

PLAY A LONG GAME

Build relationships for the long term: you may not hire someone now, but you might later. This applies to people you encounter generally, but, as mentioned previously, it also applies to people who drop out of your process or reject your offer. I met a great friend when the company she was at tried to recruit me! I didn't even interview, but we ended up becoming friends, and who knows? Maybe we'll work together one day.[5]

Helping people find what's right for them now, even if it's not where you are right now, means they are more likely to come back when it's right for them *later*.

ADVERTISING

The good news is that many industry communities have worked hard to build an inclusive space and have diverse demographics. I believe these places are some of the most effective to connect with and recruit from, especially if you can showcase a diverse team. However, not all industry communities focus on inclusive spaces and diverse demographics, so you will need to consider where you are placing job ads. If you are looking to recruit under-indexed folk, skip Hacker News and look for places where under-indexed people are more likely to be (it turns out Stack Overflow is surprisingly popular).[6]

You *can* use affinity groups to advertise your job postings. Historically, I have not found these to be particularly useful when it comes to hiring more experienced engineers, though. The more senior people get, the more selective they can be, and the less likely it is that they will find their next opportunity in a generalized affinity space—that is, a senior Android engineer will expect to find their next role via the Android communities they are part of, not a "women in tech" community. Sometimes people think that posting on these boards will be sufficient and are surprised when it's not, so it's important to collect information such that you can know whether or not it's working and adjust your approach if necessary.

If you start evaluating inclusion and representation as you evaluate how to spend your recruitment budget, you'll likely make different choices on how to spend it. It's worth tracking applicant sources against diversity metrics and using that as a factor in what and when to invest in using that source. By analyzing

---

5 In fact, you'll meet Jill in Section 4, "Self-Improving Teams", where she wrote one of the case studies.

6 I did not expect this, but when we did some research into what women looked for when considering a new role, this is what we found (*https://oreil.ly/Oitfx*).

sources in this way, it's possible to identify which sources generate a large, homogenous, unqualified application flow and then eliminate those sources as a poor use of both money and time.

TARGETED OUTREACH

The more senior the candidates you are looking for, the more you can expect to have to reach out directly to people. My observation is that women in senior roles are likely to go and work for someone *they* know.

Work on your own network, and make yourself available. Follow more under-indexed folk on social media even after you discover they are not just offering an education but are being normal human beings with varied interests. Make an effort to be more involved in communities where there is better diversity and more effort for inclusion. For example, choose "welcoming JavaScript evening with soft drinks and childcare" over the "brogramming with beer" event. Choose the Slack community with a strong—and enforced—code of conduct over the one that is "fine" except for that channel, and that one, and oh wait, is there really a channel for *that...*

It's important that your recruiters understand and prioritize diversity in their sourcing and that they are able to connect with under-indexed candidates. Diversity in your recruiting team also needs to be a priority and can really help improve conversion and experience for under-indexed candidates—especially when you listen to them about the feedback they have on your process and what they are noticing from candidates. Ideally, you have a recruiter who sources for you; if not, good luck with the art of the cold email and endless rejection.

A strong partnership with your recruiter can really help; let them tell you which candidates they feel strongly about, and make some extra time for candidate sell calls preprocess. It can feel like not a great use of time—as engineers, we are used to doing things where input leads to more concrete output. But remember, we're playing a long game—you don't know when it will pay off. It's also an opportunity to learn what kinds of questions people ask and what concerns they have, which can help you start to address those concerns up front.

Improving Evaluations

There's no point putting effort into your candidate pool if your evaluation process is unstructured and, as a result, riddled with bias. Being clear about what really predicts success helps us get clear about what we are looking for, why, and how to know when we find it.

JOB DESCRIPTIONS

Periodically, some kind of ridiculous job description goes viral on the internet. The details change, but invariably it follows a standard format: a long list of qualifications, a demand for "passion" and/or "commitment," and...a very low salary.

At the core of it is the problem with all bad job descriptions: *the people writing them think about what they need much more than what they are offering*, forgetting that hiring is a two-way endeavor.

When we hire, we often start from the problem that the person we hire is expected to solve. The more superficial the take on the problem, the worse the job description. The parent who looks for a nanny with the kinds of qualifications and commitment they have themselves is no different from the manager who, upon losing a key member of the team, writes a job description that maps to what that person did—and overlooks the reality that the person who left did not arrive doing all those things and almost certainly left for a good reason. Both of them have identified the requirement as a specific person, currently unavailable, and wrote the job description in search of the next best thing: another person who is basically the same but willing and available to do that job.

This is an extreme, of course, and I'm sure you have never done this. But I bet what you *have* done is added extra lines to filter people out, changed your nice-to-haves to requirements. I bet you have iterated over your job description and neglected to step back and consider it holistically. I bet you have made decisions that made sense at the time and not had the time or data to consider them holistically.

At Automattic, the job description for standard engineer job postings (JavaScript/backend) included a question to submit a response from making an API call. Easy to do using cURL (*https://oreil.ly/TBXVM*) or whatever. It was added thoughtfully and rationally, and it was intended to be a very quick filter of people applying to these jobs who did not know how to make a standard network request (which is typically a very minimal requirement for any developer).

The thing is, this filter also filtered out everyone who found that question pretty insulting. It turned out, this was a lot of people. And particularly when we looked into how women were experiencing the job postings, they called it out as something that gave a bad impression and considered it ridiculous to ask.[7] Which, to be fair, it was. The vast majority of developers—including even the most inexperienced developer, a developer barely out of the shortest code-school program—have made an API call. This specific filter was implemented in response to the volume of particularly low-qualified inbound by engineers trying to run hiring with no recruiter, and as time passed and things changed, it did not work as intended.

When it comes to job descriptions, a commonly cited fact is that women and people of color feel they need to meet more of the requirements to bother applying (a rational assessment in the biased world we live in).[8] The more requirements there are, the more people you put off, so think carefully about what you want to add.

When you make the recommended changes to your job descriptions to make them more appealing to under-indexed candidates, you may indeed get more applications from qualified under-indexed people. The thing is, you will get *even more* applications from unqualified people.

And then you'll want to filter those people out—understandably—and you'll start adding things in. But not all those filters will work as intended, and eventually, your job posting will stop making sense.

The difference between a poor job posting and a great job posting is this: the poor job posting attempts to be a filter. To produce something akin to a regex for people: match these things, go forward. Do not, move on.

A great posting is not a filter; it is an invitation. It is an invitation you write to the kind of person who will add to your team, where you explain what it is they will add and what they will get in return, in a way that goes beyond the exchange we make of labor for money under capitalism. It should paint a picture of the kind of work they will do, and how they will do it, and why it will matter. Tell them—honestly but aspirationally—about the current moment, but also tell them what kind of potential is to come.

Some people will read that invitation and think it's for them when it's not, and that's OK. This is why you do resume reviews. Don't write your job

---

7 "We removed all the little games from our job posting page. We were trying to test people's attention to the job posting and filter out unmotivated candidates; it turned out we were also putting people off who we want to apply." (See my blog post "Sharing the Data: How Technical Women Navigate Their Career" (*https://oreil.ly/kE7i_*).)

8 It turns out that this take is more nuanced than "women need more confidence" but is about being respectful of time, not setting themselves up for failure, and...believing the rules as written are the real rules that people follow. (See Tara Sophia Mohr, "Why Women Don't Apply for Jobs Unless They're 100% Qualified" (*https://oreil.ly/IYO8A*), *Harvard Business Review* (August 25, 2014).

descriptions to turn unqualified people away—it will be hard to deter them anyway. *Write your job descriptions to invite the people you want in.*

If you decide to iterate on your job postings, here's your 10-step plan:

1. Look at job postings from other companies that you like and ask yourself what about them is great. How can you emulate that?

2. Consider how bad your job postings are. Do they need to go in the bin entirely, or can you salvage something?

3. List the competencies someone needs to be successful.

4. Rethink that list and cross out anything that someone could easily learn in your current environment.

5. Write a list of what someone will get out of your environment, what they can expect to learn, and why their work will matter.

6. Review your lists with relevant people and add or adjust based on their feedback.

7. Turn it into a first draft of your job posting.

8. Run it through a service, such as textio (*https://textio.com*), to make sure the posting uses inclusive language and isn't off-putting to under-indexed groups.

9. Send it to a couple of people you would love to hire, and ask them for feedback—would they apply for this job? Is there anything about it that's off-putting to them? What would they expect the job to be? Iterate based on their feedback.

10. Measure success based on conversion from outreach emails or number of qualified applicants. Do **not** base success on inbound application volume.

RESUME REVIEW

When I had the misfortune of running hiring without a recruiter, I had to review a *lot* of resumes. It was not fun. I would sooner fight with Dagger[9] or run a database migration. Unfortunately, those things weren't my job anymore because I was A Manager, so I reviewed resumes and (quietly) regretted my life choices.

---

9 Android dependency injection framework, gradually being replaced by Hilt, which I'm sure will also be more fun to fight with than resume review.

Something I quickly learned is that most resumes are terrible.[10] People write collections of buzzwords, talk about their hobbies, and add all sorts of other things. Once I saw[11] a resume that was about six pages in a 10-point font. People would attach pictures of themselves; men would talk about how much they love their wives—what is the expected response to that? "I'm glad (I guess?) that you have found love, but it has really nothing to do with whether or not you are qualified for this job."

Resumes are the professional equivalent of clicking through on the social media trending topics: often tangentially—at best—related and a full spectrum of human weirdness.

To help sort through resumes more easily, I decided to score them on a 10-point scale, with 1 being the lowest. The numbers 1–5 covered the "heck no" category and essentially captured my annoyance that I'd wasted my time looking at the resume at all. So an outsourcing company would score 1, a not-yet-graduated dude applying for a senior engineer position might get a 3 (but probably a 2), and someone who clearly did not meet the requirements but was at least in the right domain would get a 5. Above 5, it got more interesting:

- 6: probably doesn't meet requirements, meh overall (weak no)
- 7: may not meet requirements, but something interesting about them (weak yes)
- 8: clearly qualified (strong yes)
- 9: clearly qualified and clearly excited about this job (strong yes)

In my unscientific analysis of my completely subjective method, I came to the following conclusions:

- 6: don't waste time interviewing
- 7: panning for gold; most people didn't make it, but those who did were some of the best hires I ever made
- 8: most likely to drop because of a competing offer
- 9: almost always hired

---

10 My suspicion is that most resume-writing advice is not written by people who review a lot of resumes.

11 I can't claim to have read it; all I can say is that I tried.

Because I both reviewed every resume and interviewed every person, I had a very short feedback loop. The experiment of interviewing 6s did not last for long because I was clearly wasting my own time. It also gave me clarity on when to prioritize: time spent on 9s was clearly the most worthwhile, so I would be more willing to make time there, schedule an interview or two more than I really had time for that week, or be more flexible.

As much as reviewing every resume was an experience I did not enjoy and would strongly prefer not to repeat, once I started managing recruiters, I *really* understood how important calibration was. While I still think on the 6–9 scale, especially in a domain (native applications) that I'm an expert in, it has become more and more important to try to codify that for the recruiters I work with.

When resume review is done poorly, it breaks the entire process. The easiest way for resume review to be done poorly is when it's too lenient. The recruiter has a different impression of "something interesting" than I do and moves someone forward who shouldn't be. For example, when hiring native application engineers, I have looked for people who have worked on complex applications. Recruiters sometimes equate that complexity to a brand name, but plenty of brand name applications are just CRUD[12] and are not that interesting technically. When working with a recruiter, you need to spend time helping them understand the space you are recruiting for and calibrating them on what you expect. Good recruiters know this and will ask you to do it, while less experienced ones may not, so when you realize you are not aligned, you will need to spend time correcting that.

That being said, resume review—especially for companies that hire internationally—is a minefield of bias and subjectivity. First and foremost, differences in cultural norms have a huge effect on what is included. There's a lot of variation in the amount of personal information people include and how people represent their work, ranging from overly humble to exhaustingly braggy. Second, cultural context plays a role. What is considered a strong resume in a place with a lot of opportunity versus what is considered a strong resume in a historically outsourcing country, where people have much less opportunity?

---

12 Create, read, update, and delete.

Third, your own emotional context while reviewing resumes can be a factor. Even judges make different decisions right before lunch, and there is no question that we make different judgment calls depending on how we feel.[13]

The answer to these myriad issues around resume review is quantification and calibration:

Quantification

Determine some kind of review scale so that you can compare performance later in the process. The level of fidelity varies depending on how many people are involved and how complex the field is. My 9-point system worked fine for just me, but when I am working with multiple recruiters, it's more useful to have a list of characteristics that we want to see on the resume, expect some minimum amount, and have the recruiter dig in further during screening.

Calibration

Every new person reviewing resumes needs to be carefully calibrated, and that calibration needs to be *maintained*. Use the same scoring system and discuss differences. The other useful calibration tool we use is "surprising failures."[14] Every so often, the recruiters pull a list of people who, based on their resumes, the recruiters thought would be more successful in the process than they were. Then we give feedback on the resumes versus what was uncovered in the process. Or, if there's reason to believe we may have made the wrong call, we revisit the person overall.

Having a good resume-review process is key to identifying a weak applicant pool as early as possible. If your interest level is low, then your option set will be small. It is harder to calibrate from a small inbound applicant pool, and especially for a new role, you may have to interview a lot more people than you want to as you figure out what the market is like. Talking to interested people will help you understand the roles, expectations, and motivations of your candidates

---

13 "Judges granted 65% of requests they heard at the beginning of the day's session and almost none at the end. Right after a snack break, approvals jumped back to 65% again" (Kurt Kleiner, "Lunchtime Leniency: Judges' Rulings Are Harsher When They Are Hungrier" (*https://oreil.ly/4LoVx*), *Scientific American* (September 1, 2011).

14 Inspired by Alan Eustace, former senior VP at Google who, when he was concerned about bias against women in hiring early on, started going through hiring packets to understand how decisions were being made and whether they were correct.

better while giving you a clearer window into what you, the hiring manager, can expect.

If you think your resume-review process needs to be updated, these are your next steps:

1. Review your rubric and how you decide which resumes are suitable to progress. Encode those expectations, and apply them consistently.

2. Set up your data so that you can see how resume review predicts performance in the process.

3. Determine your ongoing calibration strategies:

    a. How will you calibrate new people?

    b. Schedule regular reviews of surprising failures.

EVALUATIONS

It's quite possible that the skills that were prized or evaluated when you were hired are not the skills that are most needed in your role today. It's almost certain that you've learned and grown since then anyway (and if not, maybe that's the problem you should be looking to solve).

We tend to either overvalue our own skills or take them for granted. Spend some time thinking about what makes different people effective and about the hidden work that goes into that. Learn to value and articulate specific actions that lead to specific outcomes.

Whatever the case when you were hired, what is needed for success today? Think about what you want to add to the team and what your current process selects for. Invariably, there is a gap to fill, and the size of the gap implies what you do next. Minor gaps have minor fixes; major gaps require a major rethink.

As dispassionately as possible, consider what success looks like in this role and the kind of impact a strong performer could have. Try to differentiate between key skills and learned behaviors. For example, when hiring a senior engineer into a complex codebase, their ability to grapple with the complexity is not negotiable. Being able to use Git effectively is key to being successful but very much learnable on the job, so don't rule people out on the basis that they have been using another system of version control.

Competencies

The first step in hiring-process design is to identify what competencies are required for someone to be successful in the role. Ideally, you did at least some of this work when putting together your job description.

If you have well-defined engineering levels, these are a great place to start. If you don't, start making a list of what it means to be successful in the role you are hiring for. Some of the things you may want to evaluate at a high level are:

- Communication skills
- Ability to break down projects and deliver incrementally
- Ability to work with existing code
- Ability to design new features
- Ability to thoughtfully reason about architecture and trade-offs
- Understanding of and experience with testing
- How they respond to feedback

When working from defined levels, I like to make a spreadsheet that lists out all competencies for a given level (grouped as they typically are in your levels: project management, technical expertise, leadership, etc.) and then convert them one by one to the external evaluation point.

For example, for technical expertise I might include the following competencies:

- Identifies the most important/risky parts of problems and figures out how to derisk them
- Breaks down problems in order to validate and deliver incrementally/concurrently
- Is able to shard a complex problem, onboard other developers, and work collaboratively
- Can make incremental progress improving a complex system
- Can build a testing strategy that makes sense given constraints (complexity, time, likely churn)
- Can write clean, idiomatic code
- Is familiar with and makes effective use of standard design patterns

Then, I identify which parts of the process should evaluate each point. This produces an "evaluation matrix," and I can see that (1) everything is covered and (2) things are reasonably distributed.

You can also use this process to identify gaps in your current process: where some aspects are being overevaluated in too many places and others are being left out. I've put together an example of this in Table 8-1, where you can see that people's ability to work with existing code is not being evaluated.[15]

Table 8-1. Sample evaluation matrix showing what core competencies are being evaluated in which part of the process

Competency	Interview 1: Behavioral	Interview 2: Systems design	Code test
Communication skills	✓		
Ability to break down projects and deliver incrementally			✓
Ability to work with existing code			
Ability to thoughtfully design new features		✓	
Ability to thoughtfully reason about architecture and trade-offs		✓	
Understanding and experience with testing			✓
How they respond to feedback	✓		

I find the evaluation matrix the right fidelity to get buy-in to significant process change. It encourages people looking at it to consider it *structurally* rather than piece by piece. The discussion can be framed around: what are we evaluating and where are we evaluating it? Agreeing on that is best before going into the details of *how*.

Once we have agreed what we are evaluating where, we can take the evaluation points and design the stages around them. Even better, because we have clarity on what we are evaluating, we can tell the candidate what we are evaluating. It's much better to tell the candidate what you're evaluating them for explicitly because when you leave out information, you're effectively evaluating

---

15 Yes, this hypothetical team is having a lot of arguments about needing to refactor/rewrite things.

their ability to guess what you want (or whether they have someone on the inside who tells them).

It's important to select for behaviors rather than characteristics. For example, responding well to feedback is a key attribute for a strong hire as it's a strong predictor of coachability (we talked about coachability in Chapter 3). We can look for a characteristic commonly associated with this ("This person has a graduate degree, so...") or, better, we can test the behavior (give them feedback and see how they respond).

Let's take a specific example.

I inherited a hiring process featuring an exercise that asked people to write about how they would run a fictional project inside a different organization than the one they worked in currently (or most recently). There were three significant problems with this:

- The problem was very domain specific, and most people did not have that domain knowledge, meaning that many people either spent a lot of time researching the domain or missed the complexity of the domain and couldn't produce a reasonable document.

- People had no idea how this new organization worked, and their view of how to run a project was heavily shaped by their previous work experience, in nonobvious ways. Their plans weren't aligned with *this* organization (and understandably so).

- The rubric was derived from the assignment and assessed whether it was a reasonable approach to the fictional project. Which meant that overall, too much of someone's performance on the assignment was determined by their ability to guess the correct answer—a product of luck rather than skill.

We went through the process to create the evaluation matrix and derive a new assignment. We kept the same high-level concepts, in that it was focused on how to run a project, but now we asked them to write about a real project they had had significant ownership over. This addressed several of the previous issues:

- No more guessing what we wanted; instead, people could explain the problem and how they approached it.

- A more equitable time commitment: starting from people's existing experience removed the head start that people already familiar with the domain had previously had.

- Less bias from judging people's decisions in a space we understood deeply; instead, they were evaluated on whether they explained the space and their decisions such that we could understand them.

- If someone worked in a dysfunctional or overly political environment, that was clear and could be factored out more easily.

- Because people wrote about different projects, they were easier to differentiate, making reviewing them a better experience and setting them up for a better experience at an interview.

Because we derived from our leveling rubric, it turned out this exercise was suitable for all roles and was made generic, reducing the overhead for each individual role-hiring process, making expectations across the organization more standard, and making it possible for people to participate in hiring cross-functionally.

Once we had used this new assignment for a while and knew it was a strong predictor of success, we were able to revisit it and cut it down to the most useful core. Although it was slower overall, making this a two-step process of (1) evaluating the right thing and (2) making it more efficient helped improve buy-in and confidence. You can't necessarily change everything at once—and this is why you develop a strategy.

Because everything is tied to the leveling framework, if there's any ambiguity about someone's level at the end of the process, we can map the person's demonstrated competencies to our leveling framework and do a "level assessment." This can be a useful tool for resolving discussions about whether someone is correctly leveled; it removes subjectivity and "I know it when I see it" bias, and creates—or doesn't—a proper justification.

Areas to develop

Everyone has areas of growth, and if you have not identified them in your hiring process, all that means is that you do not (yet) know what they are. If they join your team, you will find out eventually.

As a result (and having learned this the hard way more than once), the thing that I am most fanatical about evaluating in any hiring process is someone's ability to respond to feedback. My teams build highly complex pieces of software for the long term—not identical widgets—and as such, change and growth are constant. Because I'm hiring people for the long term, I know that however skilled and effective they are today, more and different things will be required of them in the future, and their ability to respond to feedback (and general self-awareness) tells me a lot about whether they will be able to get there.

One mistake that stems from the (ridiculous) concept of the hiring bar is to require more and more from people and to eliminate anyone at the first misstep. This is how we "raise" the bar. By expecting areas of development and prioritizing giving feedback, we can invert this. Identifying where there is a lack of clarity about whether someone has the competency, or something that was missing, is a win. Then you can give the feedback to the person and see what they do with that feedback.

Beyond that, it's important to ask in any hiring process, "What can we realistically expect someone to learn?" For instance, you still occasionally find software engineers in the wild who are not wholly familiar with Git, often because their workplace uses something else. If they mess up a rebase, does it matter? Any capable software engineer who works with Git for a month or so will surely figure it out. Why select for it and rule people out unnecessarily?

Similarly, while working on large and somewhat esoteric pieces of software, we expect that everyone will learn the nuances and gotchas after they start working on it. I remember when I was at Automattic, someone reviewed a candidate's work and criticized them for not realizing that there were three text editors in the app; they thought this should cause us to reject them. I countered that no reasonable person would expect there to be three text editors in the app and that we should tell them that and see what happened.[16] At DuckDuckGo, I see similar issues with the details of content blocking[17] on iOS. Whatever your engineers work with day in and day out will seem normal to them. When someone who is not immersed in that misses something, it's worth asking the question "Would a reasonable person expect this?" and operating accordingly.

By using structured evaluations and some form of level assessment, you can make informed opinions about where someone's gaps are and whether those are things they can reasonably expect to close. In the previous project management example given, that company is heavily weighing project management expertise, and if you're familiar with the market, you'll know that not every developer gets the opportunity to practice that. We might, through this process, ascertain that someone does not have the experience in project management to meet the

---

16 I know, you want to know why there were three text editors in the app. The short explanation is *shrug* legacy software.

17 This is how trackers get blocked on the Apple platforms: via the content-blocking framework, which compiles a bunch of trigger/action rules.

expectations of the role and level. Then, we can decide what to do with that. If they are strong in other areas, we're in a position to support them, and once we have verified that they respond well to feedback, then we can make a case to hire them anyway (perhaps at a level below) and be clear with them what that means.

Optimizing for a fast, respectful no

Bad hires are expensive, draining, and frustrating. No one wants to make them, even though we all will occasionally. As a result, people often seek certainty by asking for more and more from candidates—missing the bigger picture of who it puts off from applying at all and how much work it creates for everyone involved.

No one likes having their time wasted, and it sucks for everyone to reject someone at the end of a process for something you knew (or should have known) in the beginning. This is why I suggest considering how to optimize for a fast, respectful no.

The important thing is that we make these decisions based on data, rather than feeling, and encode them in the process. Here are some examples—note that these are all situation specific and aspects that in different environments may not matter as much (or at all):

- In some roles, I have stopped interviewing people with mainly contracting/consultancy experience because it was clear that if people work on smaller apps for shorter periods of time, they cannot demonstrate the technical depth or level of project management we look for.

  Because of global inequity, it might make sense to talk to people in historically outsourcing countries with this kind of experience but pass on people in countries with significant opportunities for product development.

- In remote environments, there's typically a priority placed on written communication, which is required to be effective. In this type of environment, if someone cannot communicate coherently in writing, it makes sense to say no earlier (ideally at application).

  This doesn't mean someone must have perfect English. But for example, when asking people to answer some questions with their submission, I have found poorly written answers there a strong predictor of poor performance in later stages (which require longer written answers).

This is where ongoing calibration at every stage in the process is so helpful. By looking at the data, common reasons for rejection, and mapping that to things identified earlier in the process, *while also consciously and proactively managing bias and whether it is something we want to evaluate,* we can make the process more efficient and be less likely to waste people's time.

The rule to keep in mind is that when you put someone through your process, you ask for their time. The respectful thing to do is to ask for time proportional to likelihood of success. The more time you ask from someone, the higher the confidence you should have that they will be successful. Not making a decision on a meh technical interview but sending someone to code test anyway rarely pays off and is a waste of their time. It might feel like you're being "nice," but you're not being *kind.*

Stepping back for a more holistic review can be really helpful here. For instance, in a multistep process, it's worth looking at all the data points you have at some interim step in the process and considering holistically whether the candidate is likely to be successful. I don't want to reject someone based on a not-amazing interview alone because interviews are one of the most subjective parts of the process. But if we look at someone holistically considering two interviews or an interview and a document, and we think it's unlikely they will pass the next step based on multiple data points, we don't want to waste more of their time.

ASSESSING TRAJECTORY

Maybe you remember in Section 1, "DRIing Your Career" that we talked about five years of experience versus the same year of experience...five times over. This is the hiring manager's perspective on that, where you try to understand what's behind the line items on the resume. What did the person accomplish? What did they learn?

This whole concept is fraught with bias. For instance, how do you distinguish between lack of drive and lack of opportunity? How much can you really infer from tenure? What does it tell you if they didn't get promoted? What does it tell you if they did? This often says more about the organization than the individual.

Despite that, I still consider trajectory a useful thing to understand about any potential hire.[18] In particular, I want to understand the following:

---

18 The idea of where people go with what they have is the core of Adam Grant's book *Hidden Potential: The Science of Achieving Greater Things* (Penguin): "Potential is not a matter of where you start, but of how far you travel. We need to focus less on starting points and more on distance traveled." You can find a summary of core ideas in the book online (*https://oreil.ly/aj7Ci*).

How did their responsibilities compare to the job title?

Are they realistic in what they are looking for? If they think they are qualified for a specific level but their experience doesn't actually back that up, even if they take a lower-level position, they may come in with a degree of dissonance that makes it hard to coach them.

What kinds of environments have they worked in (e.g., big companies, small companies, tech companies, outsourcing companies...)?

If they have predominantly worked in one kind of company, and that's different from what they would be joining—for instance, switching from outsourcing to product, tech company to nontech company, huge company to smaller one—that can be a learning curve.

What is the complexity of the things they have worked on or led, and how has that changed over time?

This doesn't mean ruling out people who have had nonstandard or unusual paths, but it's worth understanding why that is and what drives them.

If they have held leadership roles, have they stayed in those roles long enough to make changes and see their impact?

This is critical for making that experience credible.

One problem that bias creates is that we are inclined to give some people the benefit of the doubt and not others—for instance, research from Professor Kelly Shue of Yale has shown that men are evaluated on potential and women are evaluated on past experience.[19] We could improve equity by evaluating everyone on potential,[20] *or* we could improve equity by evaluating everyone on past experience. Be conscious when you are evaluating the trajectory that you are using it to understand the quality of people's past experience so that you can more equitably evaluate them.

Better Interviews

The internet is full of Opinions on how to do technical interviews. It's quite overwhelming. Ultimately, we'll never get the information to determine what is—or isn't—working from some think piece because (1) we don't know what the

---

19 Kelly Shue, "Women Aren't Promoted Because Managers Underestimate Their Potential" (*https://oreil.ly/HYHxd*), *Yale Insights* (September 17, 2021).

20 I'm not entirely sure this would work as it seems subjective and some part wishful thinking.

candidate's experience was, (2) we don't know what their success rate actually is (and that would be muddied by other variables anyway), and (3) we can't A/B test humans.[21]

So every different opinion on interviewing is just that: what someone has tried, based on their own inclinations, experiences, and biases. And what they think is working, based on their own inclinations, experiences, biases, and environment.

The environment piece is important. A technical interview is just one piece of a process, which determines what is needed in the rest of that process. And that process itself is supposed to map to the environment someone will work in.

What I take from this is two things:

- Multiple types of interviews can work, if carefully considered as part of the overall process and done effectively.

- Maintain a healthy skepticism about companies that aim to allow you to outsource a chunk of your hiring funnel.

TYPES OF INTERVIEWS AND WHAT TO EVALUATE IN AN INTERVIEW

Like every other part of the process, an interview is looking for signals as to whether (or not) a person can be effective in the environment you would hire them into. Next, let's cover two main types of interviews (behavioral and technical), then discuss the importance of take-home assignments and other considerations.

Behavioral interviews

A high-level behavioral interview might focus on:

- How they make decisions

- How they think about architecture and trade-offs

- What best practices they consider important and how strongly they hold them

- How they interact with other people

- What they are looking for in their next job

---

21 I know, I keep complaining about this. It is probably for the best that A/B testing humans is not possible.

A behavioral interview aims to evaluate "Will this person be successful on the team?" by mapping their preferences and ways of working to the needs of the team. Someone open and flexible may be great for a startup but may struggle to work on regulated software (e.g., in finance). Someone more rigid and diligent may work well on something where change has to be slow and careful (like low-level APIs) but struggle with a more "move fast and break things" environment. It's important here that you ask questions about their actual experience over hypotheticals. Hypothetical questions get hypothetical answers, and you'll learn a lot more about the candidate if you better understand their actual experience.

Actually talking to the person can help round out the other data points you have for them. As such, it can be less about what is objectively "bad" or "good" and more about what is "necessary here" and "difficult here." Making it a two-way conversation can be helpful for assessing someone's flexibility and can give them a more accurate impression of what the work will be like. For instance, before you reject someone as "religious about TDD,"[22] it's worth asking, "You seem to feel very strongly about TDD; how do you think that would work in a more scrappy environment where we don't yet have product-market fit?" Although the answer may often be the one you expect, the occasions where it isn't will likely be very valuable.

The single most useful question I ask in any behavioral interview is "What are you looking for in your next job?" Typically, it gives me either valuable information on what to use to sell the candidate or confirmation that this is not the right environment for them.

The most common mistake I see from candidates in response to this question is what they *don't* want to deal with anymore at their current job. Often, this is very fair and understandable, but until they understand what they *do* want, it's unlikely that they will find a job they are happy with overall.

Coding and technical interviews

A coding interview aims to evaluate someone's technical competence relative to the needs of the team. A coding interview might focus on:

- How do they approach problems?
  - Do they take time to think first?

---

22 Test-driven design (*https://oreil.ly/QPMPJ*).

— Do they make sure they understand the problem, or do they make assumptions?

— Can they discuss trade-offs? Consider how different inputs would change things?

- How do they respond to feedback or questions?

— Can they look at their own work objectively and see problems?

— Do they listen to feedback?

— How do they respond to new information?

- Do they understand the components they are working with?

— Do they understand the pros and limitations of different architectures?

— Do they understand what the benefits of different data structures are?[23]

Technical problems can help answer all these questions. I had a particular programming problem that I asked about for years—it was one that involved sorting, hash tables,[24] and a problem that seemed straightforward but could be approached in different ways or built on. It is possible to get a strong signal on all the preceding questions through such a question but only if it's a supportive dialogue and not a hostile interrogation.

The core mistake of such interviews is asking that kind of question from a mindset of "gotcha," where the only correct answer is to write it down perfectly without further input. Then the interviewer learns nothing about the candidate other than they got the answer to that question right, which may be a specific, narrow predictor of raw intellect, or possibly the candidate found that specific question online, but unlikely much more.

Choosing good technical interview questions

There are three criteria for a good interview question:

- Gives a sense of problem-solving and understanding
- Is explorable and extendable
- Is deeply understood by the interviewer

---

23 Understanding is different from expecting someone to code their own version; data structures are foundational, and different ones are suitable for different problems.

24 The answer is always hash table.

Problem-solving and understanding

The problem presented needs to be decomposed into smaller problems in order to be solved. There are a variety of different answers.

This criteria eliminates:

Knowledge questions and factoids
> If a question requires memory of a minor detail that rarely arises in day-to-day programming, it is a bad question.

Answers that are easily searchable
> Any question where there might be a specific answer on Stack Overflow is a bad question.

Explorable and extendable

A good question needs to cater to the range of scenarios in a normal distribution—from people struggling to produce an answer to people who quickly produce a perfect answer, with most people falling in the middle.

One approach you can take is to ask a huge question up front and just expect hardly anyone to get to the end of it. I'm not a big fan of this approach because it means almost every candidate leaves berating themselves for not getting to the end of it, creating a bad impression of the experience.

An alternative is to build a question up and change constraints. When you build a question up, the initial question is part of the problem that is then situated within a bigger problem. Ask "How does this change things?" When you change the constraints, the problem is a manageable size, and once the candidate has solved it, the interviewer can change either the constraints or the requirements.

Whichever approach is taken, as an interviewer you should be able to see the candidate effectively reasoning about what changes and what doesn't as the situation changes, and showing intentional consideration of when code can be reused...and when it shouldn't be.

Deeply understood by the interviewer

When interviewers don't deeply understand the question, they risk assessing the candidate against the answer they "know" rather than allowing the candidate to explore it and finding out what the *candidate* knows.

It's easy to think that if two people ask the same question, they give close to the same interview. I don't believe this is true. Technical interviews are not standardized tests; they are a system involving the interviewer, the candidate, the environment, the programming language, and the question.

Even when the interviewer, the question, and the environment are constant, with a question of any complexity, it's going to be different each time because different candidates will see the problem differently (possibly influenced by programming language) and take a different route through it.

This is why the interviewer's understanding of the question is so important. You have to be able to follow fundamentally different implementations, often in a variety of languages. *The question is an island the candidate landed on for the first time. The interviewer needs to have a map, to help guide them through.*

It's obvious that solving a problem in Haskell is fundamentally different from solving it in C. But solving a problem in Python is different from solving it in Java, and both of these are very different from solving it in Ruby. Languages don't just differ in syntax but also in the library methods available, the ease of using them (see sorting numbers in JavaScript[25]), and the choice of data structures available in core libraries.

This means that you can't compare progress. If someone writing Python got through 80% of the question and someone writing C got through 70%, you can't conclude that the Python programmer was better than the C programmer. *They were doing different things.*

One solution to this is that you constrain everyone to the same language. But if someone is writing code in the language they use every day, and the other person is not, then once again...they are doing different things.

All of this comes back to what you're trying to evaluate and why. If you're hiring someone to do low-level hardware programming, it may well be important that you validate that they can competently write code in C. However, if you're hiring someone to work on a backend or infrastructure team, do they really need to be most familiar with the language you're working in? Or is it something they would quickly adapt to on the job?

TAKE-HOME ASSIGNMENTS

An alternative to technical interviews is to have the candidate write some code and then the technical interview dives into the work they have actually done. You could also do this with open source software (OSS), but it biases to candidates

---

25 Defaults to alphabetical have to be overridden (*https://oreil.ly/SXTLP*), fun times.

who have such OSS work, and it creates additional work and additional risk of bias for interviewers.

In the end, the goal is normally to answer a similar set of questions, just with a bigger, more "realistic" problem rather than a small contrived one.

More visible in take-home assignments are:

- Can they work with legacy code, or will they insist on rewriting it all?
- Can they prioritize?

Depending on the level of the candidate, the expectations change. In a more junior engineer, the main thing to look for is sound base reasoning and responsiveness to feedback. In a more senior engineer, you would typically look for a more holistic understanding of the constraints, better prioritization of problems, deeper competency with their main platform and programming in general, and good response to questioning—explaining their thinking well.

ADDITIONAL CONSIDERATIONS

It's worth considering what questions remain unanswered in technically focused interviews (or why some insight into behaviors is important):

- How do they work on a codebase of significant scope?
- How do they function in the last 20% of shipping a project?
- Do they give thoughtful code reviews?
- Will they be a good mentor to other developers?
- Will they take direction? (This is partially covered in interviews but more in a "may show up if they really won't" way).
- Will they adapt to the processes you have (e.g., code style, testing standards, automation)?

INTERVIEWING EFFECTIVELY

Having discussed what we want to evaluate during the interview, now it's worth talking about how to go about it. Remember, it's a *hiring* process, not a *hazing* process. Making people stressed out and miserable creates noise when trying to predict job performance. Hopefully, your workplace is not unduly stressful, and people treat their coworkers with respect, so replicating something more akin to *that* during the interview results in a more useful signal.

Before the interview, it can be helpful to have a document that tells people what to expect during the interview or in the hiring process overall that your recruiting team can share with all candidates to help them feel prepared.

Practically prepare for the interview

Allow space in your calendar for the interview: practical, physical, and emotional.[26] Being on time with notifications paused is important to being in a productive headspace.

It's also worth allowing the candidate to do the same. If they are doing an on-site of back-to-back interviews, give them a bathroom break and make sure they have a drink. If it's on Zoom, ask them if it's still a good time, and give them space to center themselves. If they're snatching a moment during the workday to take an interview, they may need a minute to shift into a new headspace. This kind of moment is what small talk is for.

Create warmth

In any relationship with a power dynamic, the person with the most power has the most impact on the quality of that relationship—which means in this context that the burden of building a good rapport is on you, the interviewer. Coming into the interview projecting a welcoming and positive attitude toward the candidate sets them up for more success. *The Charisma Myth: How Anyone Can Master the Art and Science of Personal Magnetism* by Olivia Fox Cabane (Penguin) is a helpful book on creating warmth.

Favor a conversational style over an *interrogation* style. Explore rather than *question*. Encourage the candidate to *talk* rather than just *answer*. It's unnatural to spend an hour with someone and never reveal anything about yourself. So try and be a bit human. If they mention a place, and you went there for some reason, say so. If they pick a language you aren't familiar with, express interest and say (if true) that you've been meaning to try it. As an interviewer, you shouldn't be dominating the conversation. But if they leave feeling that they know no more about you than they did when they entered, you probably didn't seem that friendly or make your interviewee feel very comfortable.

---

26 *Hanger*—that being hungry makes you grumpy—is real.

Create a positive experience

We can give the candidate a more positive experience of the interview by doing a couple of straightforward things. First, start and end well.

As part of your opening, it's a good idea to let the candidate know what to expect rather than diving in directly. One tool I use for this is giving people the option to ask any pressing questions before we begin. Ending on a high note helps people have a better impression of the experience overall because endings are disproportionately impactful as to how we remember an experience. Leaving the candidate with a warm thank-you and clarity about next steps makes them feel better about the experience.

Another thing that makes people feel better about an experience is having some sense of control. Where there's a possibility for someone to make choices, allow them to (e.g., what programming language to use).

For technical questions, it's important to ask a question for which someone can achieve *something* even if they cannot answer it sufficiently well to move forward. This doesn't mean that it's trivially easy; it means that there are gradations of an answer. It shouldn't be all or nothing, genius or fail. Someone who is incredible needs to be able to showcase that, but someone who doesn't really have the knowledge or the experience to give even an OK answer needs to be able to achieve something.

I firmly believe that questions should be domain free where possible. This is because the interviewee is already nervous, and springing an unfamiliar domain on them is liable to induce a state of panic. The classic response to this is that it's just an example, but if it doesn't need to be tied to a specific thing, why tie it? For instance, any question that gives a video game as an example: anyone who doesn't play games is liable to be intimidated by that or need extra time to understand it.

Manage the time

Good time management means the time spent on parts of the interview is proportional to what you learn from it. It's not good time management to make a candidate sweat for 5–10 minutes looking for an off-by-one error that you could easily point out to them. But if they have fundamentally misunderstood something about the problem, it may be worth taking more time to better understand why that is (and ensure that the misunderstanding is not a result of your communication).

Remove bias

Interviews are a huge bias vector, so it's important to manage that proactively. Taking detailed notes helps a lot here (it gets easier with practice!), as does things like noting timing so that you don't have to rely on impressions of how long it took the candidate to do certain things. Once your feedback is written up, take another pass and remove conclusions that your notes or data don't support or gendered or racially biased feedback, such as deeming the candidate to "lack confidence" or suggesting they are more junior. *Whistling Vivaldi: How Stereotypes Affect Us and What We Can Do* by Claude Steele (W. W. Norton), *Delusions of Gender: How Our Minds, Society, and Neurosexism Create Difference* by Cordelia Fine (W. W. Norton), and *Women Don't Ask: Negotiation and the Gender Divide* by Linda Babcock and Sara Laschever (Princeton University Press) are all excellent books on the impact of bias.

The words *culture fit* are not useful. They're shorthand for something that rarely has a good collective meaning and often just means "I liked this person." If you can't get specific about it, don't write it down. If you get specific about it and it's a bad reason to hire someone (I once saw feedback that said someone was a good culture fit because they "liked board games and sci fi"), remove it. It doesn't matter if it's a "nice" comment—it reinforces documenting things that are not relevant and factoring them in, which exacerbates bias. What can be helpful is understanding the person's alignment with the values of the organization: if you value being data driven or biasing to action, you will want to dig into those things and understand how much those are shared by the candidate.

Your Plan to Fix Your Hiring Process

Sometimes we get to build a hiring process from scratch, and then we can do everything "properly." But most of the time, we're fixing things. We inherited a process from someone else—including our past selves—and it needs adjusting.

Changing hiring processes is challenging because people within the organization often feel threatened by that change. Some of them may take it as a sign that you don't think they are good enough and worry that they wouldn't make it through the new process. Others may resist the idea of "lowering the bar."

Hiring processes need attention and refinement as the organization evolves, and hiring well is a huge lever for team transformation. Getting it right doesn't just mean hiring good people (although that part is crucial) but doing so in a way that is time effective for you, your team, and the candidates.

For a small-scale hiring process, you can often take on the process yourself and spend some time fixing it. However, as a leader (especially a new one), you want to be careful of the optics and the perception that you're bringing in "your people" to replace the people who were there before. It can help to bring the current team along and make them part of this from the beginning. For a scaled process, you'll have no choice because you can't do it alone, so it's important to determine what your levers are—and how to use them effectively.

STEP 1: KNOW YOUR NUMBERS

This is your baseline. Look over the metrics in "Metrics" on page 162, and identify the key ones for your process right now. This is where you set up consistent reporting, which is key to measuring change.

You will need a spreadsheet. Whatever your ATS (applicant tracking system) sales rep says, the reporting will not be good enough. Consider any manual reporting steps an investment in deeply understanding your numbers, because you need to.

At this step, it is key to understand:

- The quality of your applicant pool on the metrics you care about (qualifications and diversity, most likely). Total inbound is often meaningless, so consider the number of people you move past resume review.

- The efficacy of your process, indicated by pass-through rates and the time spent on people relative to their likelihood of success. The less proportional this is, the more problems you have.

STEP 2: UNDERSTAND WHAT YOU'RE LOOKING FOR

You need to understand what it is you are looking for, represent it in your job description, and map it to every point in your process, but particularly the resume review. The further you progress people who are unlikely to be successful, the more it breaks calibration of later stages because too many poorly performing candidates start to make mediocre candidates look good.

As you look through your process, do you understand what the evaluative purpose of each step is? Consider where it fits. It's disrespectful to the candidate to say no toward the end because of something you could have easily known (or did know) at the beginning. And besides, the sooner you say no to people who can't be successful, the more time you have for people who can be.

It's critical to be mindful of bias here, though, and to not index on shortcuts riddled with bias. "Worked on something complex" is better than "FAANG experience," and "has demonstrated intellectual rigor" is better than "degree from $EliteInstitution."[27]

At this step, the key is to understand:

- What it means to be successful in this role and how it maps to the process
- How to optimize for a fast, respectful no

STEP 3: CALIBRATE, CALIBRATE, CALIBRATE

This is the most important part of the process. In a scaled process, it's not enough for one person to be great at decision making; you need a replicable process so that everyone involved can make consistent decisions across the board.

Rubrics support consistency and help standardize evaluation but are meaningless without calibrating them. Double-evaluate everything as people ramp up on hiring, but then continue to have second opinions for a range either side of the borderline. Make it normal to ask for a second opinion. It's the collective responsibility of the team to question decisions and give feedback to one another.

At this point, the key is to be clear about your checks and balances. For example:

- In two-on-one interviews, ensure consistency and feedback for interviewers.
- Normalize second opinions and have a clear escalation path for disagreement on conclusions.
- Create feedback loops with recruiters to ensure consistent calibration of resume review and recruiter screening.

STEP 4: DEBUG YOUR PROCESS

If we can accept that our process is in the midst of improvement, we can also accept that our process is probably falling short and resulting in false negatives. Two levers to help identify these are surprising failures and wildcards.

An example of a surprising failure is someone with a strong resume who doesn't make it as far in the process as we would expect. Digging into these situations—reopening them and spending more time on the person—helps us validate our process.

---

27 Facebook, Apple, Amazon, Netflix, and Google: a shorthand for big tech companies.

Wildcards might be produced by giving the recruiter (or, where applicable, the referrer) a deciding vote on someone they feel strongly about and progressing them (this doesn't apply to final decisions but to evaluative steps). Great people are often "spiky"—that is, great in some ways but not consistent. If someone doesn't excel in some early part of the process but the recruiter is really impressed by them, it can be worth spending a little more time on the person to learn more about them. Especially in the case of a referral, real-world experience working with someone gives us the signal the interview process is designed to look for, so if they're willing to advocate for someone, it's usually worth progressing them one extra step. Sometimes you'll make a great hire that you could have missed out on, and sometimes you'll spend a bit more time to confirm a no, but showing your recruiter or referrer that you value their opinion is at least a win for that relationship.

At this point, you want to understand:

- What are the triggers to investigate potential false negatives?
- How do you—judiciously—use discretion in your process?

A great recruiter is an instrumental teammate in the process; they will know a lot about what a good process looks like and are great at identifying bottlenecks and inefficiencies. But however good the recruiter, they are not someone you can delegate the full problem to—as a hiring manager, you need to own what's needed for someone to be successful, what's flexible, and what's not.

STEP 5: STOP LOOKING FOR A STRAIGHT LINE

Hiring processes fail in two ways: they result in bad hires, but more insidiously, they shut out potentially great hires as you adopt an ever-narrower definition of what you're looking for and fail to differentiate between what is straightforward to learn on the job (like using Git) and what is not (like grappling with a complex legacy codebase).

The challenge of any hiring process is the need to evaluate people in a consistent and orderly way (like jumping hurdles), when the role of most knowledge workers is more ambiguous (like running through a forest). We also evaluate people at a fixed point in time but work with them for much longer, which is why evaluating people's coachability and capacity to grow is crucial.

Moishe Lettvin wrote a helpful blog post (*https://oreil.ly/y6TEN*) about how hiring is like "mapping the potato," which basically means thinking of a candidate's limits on various aspects (kind of like a 3D spatial plot).[28] I think about this metaphor a lot, and I think it goes well with the concepts in Gallup's *Now, Discover Your Strengths*. The reality is that great people are great at some things and not great at others. When you're hiring based on hurdle jumping, you can "raise the bar" by making every hurdle higher, but that doesn't necessarily result in a better outcome overall, especially when the job is closer to a distance race or a relay with the rest of the team.

At the end of a good process, you should have a lot more information about your chosen candidate than just "passed every stage." You should have a profile of the candidate, and you should know what they will need support with, but most important, you should know in what ways they are great and how that will add to your team. (If you don't, then maybe you should consider whether you've really found the person you're looking for—and what you can learn about your process from the situation you're now in.)

I've gone through this process three times now, at volumes between 10 and 100 hires per year. Invariably, it's a lot of work, but when the playbook is clear, the results are replicable. Increasingly, hiring well is a key competency for a manager—especially in a situation as competitive as hiring software developers.

---

28 Lettvin presents the idea of this morphing shape of a potato as the hiring committee considers a candidate. Each interviewer "fills in the potato," highlighting different aspects.

9

Making People Successful

Back in Section 1, "DRIing Your Career", we talked about the importance of finding situations where you could grow. In Section 2, "Self-Management", we talked about understanding your job and setting yourself up for success. We approached these from the angle of needing to meet your own needs, in part by *managing up*. In this section we're going to take the other angle on this—*managing down*: how you help people understand their job and set them up for success.

We'll cover one of the most powerful levers for doing this: effective 1:1s. Not as a list of questions (although those can be useful) but as developing a model of what effective 1:1s look like. We'll also look at onboarding and understanding what makes a successful onboarding, as well as how to manage people through challenging periods, whether they are struggling personally or with their job.

Of course, there are those people who get into management because they are driven by power and advancement. But many managers get into it because they like helping people and want both teams and individuals to be *better*. They have *good intentions*. Although I do believe you can get a surprisingly long way as a manager by consistently showing up and giving a damn about the person as a human being, good intentions can only take you so far. In hard situations, the intentions you have will not matter in the face of the impact of failing to handle them effectively.

The worst manager I ever had failed in so many ways. While he committed to writing code on the critical path, he failed to ensure delivery to the point that the team was fired by an internal client and the product eventually shut down. Even though he had good intentions, he failed to create clarity around goals, and he failed to resolve conflict. He thought he was being a good manager because he wasn't making the mistakes that *his* previous manager had made with him. He was wrong. Maybe he didn't know what good looked like and this was the best he could do, or maybe he knew it wasn't going well but didn't want to admit it. I don't know.

Initially, I took his intentions at face value, but once I realized that he was terrible at being a manager, I was pretty mad at him. His intentions did not matter to me. It took me years to realize that his model of "not being *that* guy" was, in fact, pretty prevalent among managers still trying to figure out what their job was. I, too, had much more clarity about the things he didn't do and that he failed at than I did about the things that the better managers I had did. The most useful thing I learned in the time I worked with him was the limitation of that kind of thinking. So once I became a manager myself, I realized I needed a better mental model of what "good" management looked like to avoid making the same mistakes. I understood that I needed to take responsibility for setting up the people who reported to me for success—and a little later learned that they would then get to choose what they did with those efforts.

Whenever I get a new direct report or coaching client, I like to ask them, "Tell me about the best manager you had—what did they do for you?" It's amazing to me how often the answer to that is someone who just...did not do any harm. Who "got out of the way so we could get on with things" or "actually showed up to 1:1s and was occasionally helpful." One of the challenges of trying to be a good manager to people is they also may not know what that looks like or understand what to expect from you. They may *think* they know—like, "If I had a good manager, I would get promoted"—but there might be some steps between getting a good manager and getting promoted that the two of you need to take together, and they need to be open to that.

In one first 1:1 with a new direct report, I asked what he wanted from me, and he was very clear: he wanted me to tell him if he was not meeting expectations. That was it. A low bar, one that I was determined to vault over. Years later, we could look back and laugh about that moment, but the best thing, having worked through what it looked like for him to expect more from me, was watching him offer more to *his* directs.

Often when people believe they are good managers, it is based on the mechanics of what they do. They show up to the 1:1s, ask questions, listen carefully, give regular feedback. To some people, that consistency is very important; to others, it matters less. But if we take away the noise of people not knowing what to expect, of the bar of management being—for some—extremely low, the true measure of whether we are being a good manager *of that person, at that time,* is the impact we have on them. It's not whether they like us or whether they get the things they think they want (or deserve). It's how they grow and what they achieve. We will not succeed every time, but we can succeed more than we might otherwise if we are realistic about what the job is.

Effective 1:1s

When we talk about managing individuals, we often talk about how "essential" the 1:1 is. Sometimes a list of questions is discussed. Sometimes we talk about using "coaching" to help people "develop." The definition of *coaching* varies wildly. Sometimes it's actual coaching as defined by the International Coaching Federation,[1] sometimes it's manipulating someone to "cope better" with a situation the manager is responsible to fix, and sometimes it's asking leading questions about something the manager has already decided the answer to.

Every time I say anything about 1:1s, it seems to make some people angry, so let's get this out of the way: 1:1s are not a sacrosanct ritual; they are a mechanism for showing up consistently and building a relationship with another person. The mechanics of that are far less important than the relationship.[2] In a good relationship where there's trust and clarity, missing a 1:1 or having a short 1:1 is not a big problem. Without a good relationship, where that relationship has not yet been built, or where other channels of communication are not open, those things can undermine the effectiveness of that relationship.

For example, when Russia invaded Ukraine in 2022, I was dealing with a huge chunk of my team being in the vicinity of a war zone. It took a lot of emotional energy to support the team members who were directly affected by being close to the war zone. Those 1:1s took more energy than "regular" 1:1s, and as a result, I kept the 1:1s for those not impacted shorter and lighter to compensate. Is that ideal? No. Would I do it every week? Also no. But in that kind of situation, it's better to be realistic about my own limitations and set expectations accordingly.

Personally, I do not use a list of questions for my 1:1s. I enter every 1:1 with two questions in mind:

---

1 "ICF defines *coaching* as partnering with clients in a thought-provoking and creative process that inspires them to maximize their personal and professional potential." (International Coaching Federation, "ICF Core Competencies" (*https://oreil.ly/mCILR*).

2 An excellent book that influenced my thinking on this topic is *Leadership and Self-Deception: Getting Out of the Box* from The Arbinger Institute (Berrett-Koehler).

- What does this person need (most)?
- How do I help them get it?

Let's take the second question first. Asking *"How do I help them get it?"* instead of "How do I give it to them?" is a deliberate choice. I don't see *giving* them what they need as my responsibility; nor is it realistic or desirable. It's inherently limiting to be in a position where you can only get what your manager can give you. It might be suitable for entry-level people, maybe, but not much more than that. Part of getting comfortable with managing senior people is having value beyond "helping" or "mentoring," so to advance in leadership, you need to adjust your mindset here; otherwise, you will hold back both yourself and the people who report to you.

"What does this person need (most)?" guides my thinking about the person week by week, in the 1:1, but also in an overarching narrative.

In the actual 1:1, the conversation typically follows a pattern like the one shown in Figure 9-1; notice that this hits the components of management we discussed in Chapter 6: support (how's it going/how do you feel), direction (information/context), practical help, and feedback (development).

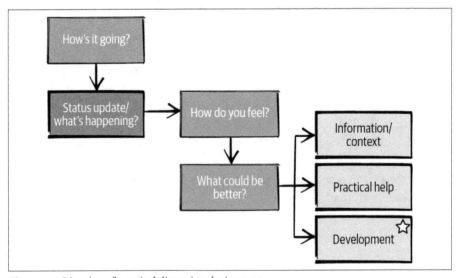

Figure 9-1. Direction of a typical discussion during a 1:1

It starts with a check-in like "How's it going?"—an open-ended question that some people respond to with a status update and some do not. The important thing is what comes next: finding out how the person *feels*. Sometimes a status update is useful context for that; sometimes it's just how we get there. Either is fine.

Then we move into the most interesting part of the 1:1: what could be better.

Providing information and context is part of direction, and it's where I spend a lot of time with new hires or someone taking on a new responsibility. They may have questions about how things work, or what they share about how they think things are going will surface things they may be missing or need to pay more attention to.

Practical help is less common but still present; if someone is overwhelmed or needs to escalate something, then that's something we would spend time on. Equally, I may help them with finding help elsewhere.

Development (part of feedback) is the most interesting part of the 1:1. This is about helping them be more effective (or sometimes just happier at work). For many people, it's not necessary to be in this space every week, but making some time for this more often than not is key to supporting people's development.

Figure 9-2 shows my model for developmental conversations.

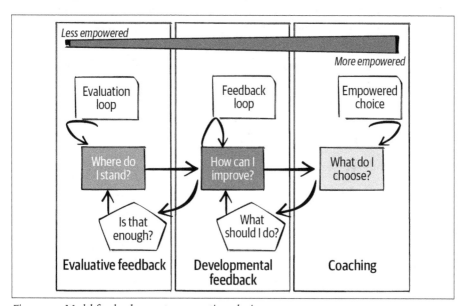

Figure 9-2. Model for development conversations during 1:1s

We talked a lot about feedback in Chapter 3, including the distinction between feedback and coaching. Coaching is where the individual decides what they want to do, and it can—in context—build on top of feedback or exist outside of it in a pure coaching relationship. Another useful concept from the book *Thanks for the Feedback* by Douglas Stone and Sheila Heen (Penguin) is distinguishing between evaluative feedback (where someone stands) and developmental feedback (how someone can improve).

Evaluative feedback is critical—if someone doesn't know where they stand, that can undermine everything else—but it is the least empowered place. The failure mode is the evaluation loop, where you keep discussing the evaluation but it doesn't change because there was no meaningful action. In this space, there's a lot of emotional work on both sides but no movement.

Moving into "How can I improve?" moves into action and as such is more empowering. The person can change the outcomes or evaluation through changing their actions. However, it's less empowering than coaching because it's more about implementing suggestions. The feedback loop is more productive than the evaluation loop, but progress is linear. The disempowerment trap is to return to evaluation through the question "Is this enough?"

Coaching is the most empowered space because it is where the person asks, "What do I want?" The empowered choice loop is where transformation—or exponential growth—happens. The disempowerment trap here is to give away the choice and ask for direction instead.

I want to be clear that no space in this model is bad; typically, we need to visit all of them. However, the difference between working with someone who is highly coachable and someone who is not is the time spent in each place.

In Chapter 3, we talked about what it looks like to be coachable. Applying it to this mode, someone highly coachable:

- Gets some evaluative feedback
- Takes suggestions well
- Figures out how they want to build on it and what they want to do

This follows the top path, and as a result, very little time is spent in evaluation, some time is spent in developmental, but the bulk of the time is in the coaching space, and that extends well beyond the time spent together.

Someone who is not coachable:

- Doesn't have ideas on how they can improve
- Asks what to do instead
- Falls into an evaluation loop

This follows the bottom path and cycles down and around—that is, even after moving into developmental, it returns to evaluative. The bulk of the time is spent in evaluation, with a small amount of time in developmental and none in coaching.

Although the evaluation space is not a hugely productive one, it's unwise to skip it entirely. Someone who is hard on themselves may seem very receptive to feedback and full of ideas about how they can improve, but if they come to this from the concern that they are falling short and need to do better, it's less empowering than if they are confident that they are doing well and are valued, and they are deciding what they can do to truly thrive.

Applying the 1:1 Model When Managing Up

If you consider the diagram in Figure 9-2, it's clear that it takes both people to get into the most empowered space. If you have a manager who actively works to keep you in the disempowered, evaluative space, this may not be a good overall work situation for you. If you can get the conversation into the middle space, by showing you're acting on things and building feedback loops, you can potentially get some useful information that can help you grow. If you know you need to get into that most empowered space, and you can't get there with your manager, then working with an external coach can help.

The Value of Complaining

This model of 1:1s does not cover complaining. I know some managers say, "Don't bring me problems; bring me solutions," but personally I don't subscribe to it; when leaders won't take complaints, it shuts down an avenue of feedback and transparency. In *Hidden Potential* (Penguin), Adam Grant says, "Weak leaders silence voice and shoot the messenger. Strong leaders welcome voice and thank the messenger. Great leaders build systems to amplify voice and elevate the messenger."

I love when people complain to me. Of course, complaining is a national pastime for the British, and we don't just limit ourselves to complaining about the weather or the poor availability of good tea when traveling. Really, your average Brit can complain about anything.[3]

But complaining can be a useful part of 1:1s as a source of rich information for a number of reasons:

It helps you suss out problems.
> First and foremost, complaining is an act of trust, one that gives you, the manager, the opportunity to address the issue and channel solutions. If people don't trust you enough to complain to, you may never learn what's wrong.

It is predicated on the complainer's experience.
> Some have the idea that feedback is not worth giving because all we ever know the truth of is our own experience.[4] When someone who reports to you tells you about their experience, there's probably some implicit feedback (as we discussed in Chapter 6) there that can help you improve as a manager. For instance, you may realize the person is complaining because they are missing some context that you should have been sharing with them—and probably should be sharing with other people, too. This is something you can then act on to improve context sharing and clarity across the team.

It shows you what they value.
> People usually don't complain about things they don't care about. So when they complain, they are showing you what's important to them. Do they care a lot about transparency? Delivery? The specifics of version control? Whatever it is, you have an opportunity to understand what matters to them and why it is annoying them so much. This can help you understand what drives and motivates them and what they need to be happy—and effective. It can also help you surface patterns across the team.

---

3 You may be thinking, "Is this why British food is so bad?" I don't know, but I wouldn't rule it out.

4 From the book *Nine Lies About Work* by Marcus Buckingham and Ashley Goodall (Harvard Business Review Press); a summary of this specific idea can be found in the *Guardian* article "Why Feedback Is Never Worthwhile" (*https://oreil.ly/ktbBT*).

It helps untangle conflict.

When two people on a team complain about each other, what you're hearing is two sides of a conflict. This is great! Now you have the information you need to help them resolve it, whether it's talking them through how their actions or communication have landed on each other or identifying the values or priorities or other nerve that the situation is hitting. Don't be a go-between, though; aim to coach the complainers through the process of building a more constructive relationship. Bonus: because they both have demonstrated that they trust you, you can try to use the transitive properties of trust to help them create some goodwill for each other, too.

It's an opportunity for coaching and clarity.

Often when people complain, it's because they think something is out of their control. This gives us the opportunity to help them grow their circle of influence. For example, if someone complains about a peer not delivering, maybe it's an opportunity to get them to see how they can be helpful or how they can give the peer some feedback. Maybe they complain about some company policy, and it's an opportunity to understand the broader reasons and implications, so you can help them to work within their constraints rather than resorting to blame.

It broadens your perspective.

We are often focused on the biggest and most urgent problems, but the minor complaints we hear today can be signs of the pressing problems of tomorrow. What can we learn from them? What can we get ahead of? Sometimes, it's just a helpful reminder that the most pressing problems are not evenly distributed across the team, and it can help us have a sense of perspective and progress.

It's an opportunity for empathy.

When we are focused on the biggest challenges of running a larger team or organization, some complaints can seem a bit like… "You're bothered by that? Really?" It might not be your biggest problem, but it is theirs, so take a deep breath and hear them out.

Remember, it's common with "nice" people to give others the benefit of the doubt, to be slow to speak up, and to play down their concerns. This can result in minor annoyances going unaddressed until they become much bigger than they ever needed to be. Turning those small, early complaints into constructive feedback can save managers a lot of work down the road.

Of course, sometimes complaining can become toxic. It's important that people both are and feel heard, but it's also important to keep things constructive. Recognize when complaining (including your own) is becoming unproductive, set some boundaries, and escalate when necessary. Asking "Do you want to vent or do you want solutions?" can be helpful to give someone space to vent and get something out of their system, but be cognizant of how much venting there is. Immediately after a big disappointment? OK. Week in, week out for months afterward? Time to redirect. Everyone gets to feel how they feel, but what they do with that is a choice: either work to change it or work to accept things the way they are. Living in a place of frustration indefinitely has no good outcome.

Onboarding

Back in the time before March 2020, when travel was a thing, I went to the Galápagos Islands. At the time, I was nomadic and spent about half my time in Colombia. I had a person who handled all my travel arrangements. For about five years, he basically ran my life, and I just followed the instructions and got on the planes. Short hop to Ecuador, a few days in Guayaquil, then another short hop to Seymour Airport, and I was there. In the Galápagos.

Now for some context: I grew up in the United Kingdom, which means that I learned basically no history or geography in school.[5]

So here I was, living in a beautiful mountain on the equator, in Medellín, also known as "the city of eternal spring." My previous remote-island experiences included Easter Island and Tuvalu, both beautiful, tropical places.

And then the plane comes in to land and...

That was the moment I learned that the Galápagos Islands are a bunch of volcanic rocks in the middle of the ocean.

You might wonder what this has to do with anything, but think about the last time you started a new job.

Maybe you had someone helping you along; maybe it wasn't too far away; maybe you expected it to be like somewhere you'd been before. But at some point, you confronted reality and learned what you didn't know you didn't know. You had—like me—about enough information to buy a plane ticket, and like

---

5 My theory: knowing too much about these things opens too many difficult questions about colonialism. Although at least we learned about evolution.

the much-studied creatures of the Galápagos, you either adapted to your environment or left it.

Onboarding is the period during which we have the biggest opportunity to help people adapt and thrive in their new environment. The first three months set expectations and perceptions in a way that gets harder and harder to change over time. But too often, onboarding is treated as a series of boxes to check rather than an on-ramp for someone to be comfortable in their role.

In my time as a manger, I've encountered quite a few people who were badly onboarded or, as I call it, "nonboarded." As a rule, it takes as long or longer to fix the bad onboarding as it would have taken to poorly onboard them. Sometimes we're able to re-onboard people and make them successful...and sometimes it's too late.

It's always baffling to me when companies do all the work to hire someone and then don't set them up to be successful. As we saw in Chapter 8, hiring takes a lot of work, and it's time-consuming on both sides. As we will cover more deeply later in this chapter, managing people who are not delivering in their role is extremely time consuming—whether you are able to get them to be effective or manage them out of the organization—not to mention emotionally draining. Onboarding people and helping them to become effective from the start is—by far—the easiest option.[6]

So what does a good onboarding experience look like?

At a high level, proper onboarding means two things:

- Creating a feeling of belonging
- Creating a sense of accomplishment

BELONGING

Feeling a sense of belonging is a human need, like food or shelter. It's feeling accepted as a part of something, such as a team or organization. It's a crucial aspect of group cohesion and psychological safety.

If someone doesn't feel like they belong, it will be detrimental to their effectiveness. How will they ask for, or offer, feedback? How will they ask questions they need to know the answers to in order to do their job?

---

6 Sea lions also have trouble with onboarding. Pups in the Galápagos have been staying with their mothers longer and longer, causing reproductive rates to decline as the mother can take care of only one pup at a time. This is causing the sea lion population to shrink. So if you want your team to scale, don't be like a sea lion.

Belonging is also the level at which it's important to think about inclusion; we want everyone to be able to feel they can belong on a team, regardless of gender, race, or sexual orientation. It's important to acknowledge that if someone is the first or the "only" on a team, that feeling of belonging may be harder to find. It's important to pay attention to inclusion issues that might impede the sense of belonging and proactively address them. As we discussed in Chapter 8, many companies spend a lot of time diversifying the hiring pipeline, but there's little point in doing so if "diverse" hires are not included and are therefore less likely to be successful.

Here are some concrete suggestions for your team to create belonging, particularly in a remote environment:

- Encourage everyone on the team to schedule a 1:1 with the new person to get to know them.
- Add an extra call to welcome them to the team, and do something social but not too awkward, like play a game.
- Share a welcome post to the entire company, and/or welcome them in the all-company/broader team call.
- Host a channel in Slack (or whatever software you use) where all new hires can ask questions and get answers.
- Some organizations have "culture buddies"—people in another part of the organization who can help them adapt and understand how things work.

A welcoming team environment became especially important during the times when people didn't know when they would actually meet their colleagues. You can't rely on them eventually finding some connection at the office or over the course of a meetup, and it's important to build it in—small and often—sooner.

ACCOMPLISHMENT

From an onboarding perspective, it is this second point (developing a sense of accomplishment) that can be harder to ensure. That's because how we define achievement is often variable across teams, from one individual to another, and most of all, because owning an achievement is often a matter of perspective for the new hire.

For engineers, achievement is normally tied to the work: the code written, the design expressed, the feature shipped. Regardless of how you define it, achievement is important because the person you hired is there to do something. You selected them based on their expertise and what you thought they would contribute to your team. The sense of accomplishment is their own confidence that they have contributed something meaningful.

Here are some concrete suggestions for your team to support a sense of accomplishment:

- Craft onboarding checklists that take a new hire through everything they need to be able to start on their job asynchronously, and make the checklists easy to refer back to as needed—just note that these checklists can seem overwhelming and impersonal and cannot completely replace human interaction.

- Take care to find a good first task and a good first project to help people get to grips with the workflow and ship something meaningful quickly. A first win helps people feel more settled. (If it's hard or slow to ship things, especially if you are hiring many people, you likely need to look at your developer experience.)

- Give every new hire an onboarding buddy who works on the same platform and project; that person is responsible for making the new hire's first project successful.

IMPACT

It's worth noting how belonging and accomplishment feed on each other. Someone who does not feel like they are contributing to the team will have a hard time feeling they belong on it. Similarly, without feeling some level of belonging, the person may find it hard to feel like they have the support to accomplish much.

As you ramp up your new hires, talk to them about these topics. Ask how connected they feel to their teammates, where they get their sense of accomplishment, and what they think they've achieved so far. Reflect back to them the support they gave their new teammates or other things they have already done that created value.

The faster you can foster a sense of belonging and accomplishment, the faster you can get your new hires to start consistently delivering on the things your team needs them to do.

IMPLEMENTATION

This all may seem like a lot of work for you as a manager, but as you scale, you will want to shift away from onboarding individuals directly and toward creating an environment of collective responsibility around onboarding:

- Collective responsibility for belonging
- Collective responsibility for accomplishment

Collective responsibility for belonging

When the team takes collective responsibility for belonging, they will make an effort to make people feel more welcome. While others may have direct responsibility (e.g., as the onboarding buddy), anyone may notice a new person struggling and look to help or intervene. Setting expectations and framing in team meetings before they arrive can be helpful, especially if it's been a while since you last added someone to the team.

Collective responsibility for accomplishment

Any codebase that has been around a while will have some gotchas in it: things that people on the team know and work around. There is also often complexity around the specific domain; for instance, if it's a photo-editing app, there will be aspects that new hires probably haven't encountered before. As new people navigate their onboarding, they will need to ramp up on the things the rest of the team is used to, and if the team feels a responsibility for helping the new person succeed, they will look to help them. For instance:

- Noticing the new person needs to touch something particularly gnarly and offering to pair on it or do a walk-through of that part of the codebase
- Offering a kind but constructive early code review
- Producing additional documentation that they realize the new person would benefit from
- Tagging appropriately scoped "nice to have" bugs and features in your ticketing systems as appropriate for onboarding

Supporting

We started with onboarding because new people often need the most support. But everyone needs support sometimes. As a manager, you will need to support your directs and, to a lesser extent, their directs, but ideally, you will build a culture where people on the team support one another.

BUILDING A CULTURE OF HELPING

The worst thing one of my engineering leader friends could say about someone is "They are not *helpful.*" There is a specific tone of disdain that unfortunately I cannot capture in writing, so you will just have to trust me that it is *damning.*

Maybe this seems disproportionate to you, so before we go further, take a minute to think about the people you're most glad to have worked with. What did they do?

Now think about the people you were glad to never work with again. Not necessarily the extremes—just the ones that, on balance, you would prefer not to work with again. What did they do? What did they *not* do?

A helping culture can be a powerful accelerant. When people get help, they spend less time stuck and feel more supported. When people *help*, they feel better about their own capabilities. In short, it's a powerful inoculation against imposter syndrome.[7]

When people don't get help, they get stuck, and they stay stuck for longer and feel more alone.

When people don't help, it can create poor incentives across the team. If someone refuses to do interviews in order to spend more time coding, and then they get promoted and their more helpful teammate does not, that incentivizes people to be individualistic rather than team minded.

Of course, sometimes helping can be a distraction: you don't want to enable the person who helps everyone else at the expense of getting their own work done. If someone is too intent on helping or helps indiscriminately, it's worth getting curious about why.

---

7 Don't confuse imposter syndrome with the pathologicalization of under-indexed people reacting to hostile environments (*https://oreil.ly/WMyUj*).

Adam Grant's research, elaborated on in the book *Give and Take: A Revolutionary Approach to Success* (Penguin) describes three groups of people. The easiest to notice are givers (the people who do all the code reviews/releases/interviews) and takers (the people who do the least of those things).[8]

The third category—matchers—is the most important. These are the people who enforce fairness. They will help people who also help them and refuse to help people who don't help. **These people create fairness in an organization.** When givers have takers around them, they burn out (as we discussed in Chapter 5), and the culture of the organization becomes individualistic. When givers have matchers around them, both the givers and the matchers thrive as they support one another.

As a manager, regardless of your personal inclinations, it's your job to be a matcher (or load balancer!) for your team in order to make it an environment where givers can thrive. This means creating opportunities for helping, encouraging helping as part of people's jobs, and rewarding helping.

Create opportunities for helping by surfacing struggle and expertise. Project checkpoints, like daily standups or technical designs or code reviews, are great for this. One of the biggest blockers people have in offering help is knowing whether it's needed or welcome, so the more you can create visibility into what's happening (without drowning people in details), the more opportunities for helping you create.

Encourage helping explicitly by creating and supporting opportunities for peer support. Encourage people to reach out to people who could help them or where they can offer help.

Reward helping by ensuring that people get credit for their actions. This might be private, like thanking someone in a 1:1, or public. On one of my previous teams, we started doing monthly "snaps"[9] posts where people would call out those who had helped them, and the concept rapidly spread across the organization.[10] At DuckDuckGo, we have a "green bow tie award," and I make a point to call out those who were nominated in the team channel each week. It's important that you make it clear that you value this behavior week in and week out as well as during the performance-review cycle.

---

8 You can find a summary of this concept in Adam Grant's TED talk (*https://oreil.ly/2Se3-*).

9 Yes, this is a *Legally Blonde* reference.

10 Without many people realizing it was a *Legally Blonde* reference.

It's important to think about helping as being something everyone does and everyone receives. Under-indexed folk may feel less comfortable asking for help, fearing they will be judged harshly for it.

Finally, think about helping as being core to leadership. As Edgar H. Schein says in *Helping: How to Offer, Give, and Receive Help* (Berrett-Koehler):

> *As leaders interact with others, they must realize that the best way to improve the organization is to create an environment of mutual help and to demonstrate their own helping skills in their dealings with others in the organization. Though it may seem counterintuitive to see one's subordinates as clients who have to be helped to succeed in their job, in fact, this is the most appropriate way to lead an organization. One way to define leadership, then, is to say that it is a process of both setting goals and helping others (subordinates) to achieve those goals.*

CHOOSE BEING KIND OVER BEING NICE

One mistake I've seen managers make again and again when someone is struggling is that they're *nice* about it and they don't address the issue: "Oh, $person is having a bad time right now; let them be." This might be a reasonable strategy for someone having a bad day, or even a bad week, but beyond that, it becomes very, *very* bad.

Here's why—at some point, one of two things happens:

- There's a performance review, and you have to tell that person they are not doing great. Whether they are surprised or not, their trajectory is damaged, and you have to deal with the fact that you did not tell them before and *did not help them.*

- They drift away from the team, the team stops expecting things from them (and resents you for not dealing with it), and they become the person who can't be relied upon. Once they're feeling better, they probably decide that they want a fresh start—on a different team or at a different company. And then you likely have to source a replacement.

I once took over a team and found a person who, after a truly traumatic series of events, had just been left to drift away from the team for *a year*. He'd become some kind of team ghost that no one expected anything from. His personal situation was heartbreaking, but his work situation made me *angry*. So many people were like, "Oh, that sounds terrible, I'm so sorry, what do you need?" Here's the thing: he did not know what he needed, and he did not know what was available to him. These people were being *nice*, but they were not being *kind*. No one had helped him.

So what happened next? I got on a call with him and did a reset. He had spent the last year in and out of work trying to handle things in his personal life, and I asked him what he thought a normal amount of time off would be for someone in his situation. Given what was going on with him, I told him that we could realistically have expected him to be out for eight or nine months. One of the problems was that he felt guilty for being out so much, and I needed to help reset his expectations for what would be considered a normal range for being out. Then, I reassured him that we were going to let the last year go and work on reintegrating him into the team going forward.

He cried. And it was apparent to me that despite all the *niceness* shown by everyone else at the company, this was the first time he'd felt actually supported at work.

It was not completely smooth, of course, and it took some time. But things got better. Eventually, he became the teammate that people had known and missed again. He stepped up and took a leadership role. Although we could never do anything about the tragedy in his personal life, we could untangle the impact on his career and get him to a place where he could enjoy his work again.

I learned a lot from this situation, but the biggest lesson I took is the corrosive effect of *niceness*: the potential damage of the boundary where someone's work is their work and their life is their life. People are who they are, and whatever they're going through during the entire 168 hours of each week doesn't stop for the 40 or so they spend on their job. In a distributed context, where people work from home, this can be even harder; going to an office might be a temporary respite, but you cannot escape what is happening in your life when your office is just another room in your home.

I'm not saying you need to get all up in the personal lives of your team—absolutely not. But, as we talked about in terms of managing burnout in Chapter 5, you will be much better able to support people when you allow them to be a full human. A measure of the relationship you have with someone is that they will tell you when they're not doing great and *let you help them*. It is much, much easier to help in the moment than after the fact.

Here are five key approaches for supporting people having a bad time:

- Reflect back the struggle.
- Reduce required emotional energy.

- Let people be.
- Agree on the plan together.
- Recognize that if you don't take the time to deal with an issue, the time takes you.

Reflect back the struggle

While working in a distributed context, I noticed that people working from home were often more isolated—and this became more broadly true during a global pandemic. Some people work remotely because it supports some kind of wild, adventurous, international life. Some people work remotely because they are introverts who hate office life. But one thing I have seen again and again is that someone who isn't doing so well withdraws: they participate less in the team and seem less present, and it's easier for this to go unnoticed when you only see someone's virtual presence and not their physical one. They go a bit into their own head or are too deep in the overwhelm to see it clearly.

In the coaching model, there's a concept called "name what's going on." It's basically self-explanatory: you name the thing, the elephant in the room, the subtext of what they are saying—or not saying. But it's been amazing to me how often I listen to someone talk through the details of where they are at and then I reflect back what I've heard overall, and this is what helps someone realize where it is they are at.

Or I reflect back to them the bigger picture of many conversations. An example.

I was managing someone who kept coming to our 1:1s and saying he was really struggling personally. He also kept saying it would be OK because something that was going to happen in the next few weeks would make things better.

Multiple such future events came and went. I think they did help him, but the bigger picture didn't change. And so one day, I reflected back to him that he had been in this cycle for a while and suggested that maybe the answer wasn't in these one-off events but rather in some change to day-to-day life. I offered him some time to make those changes, if it would be helpful.

He left that conversation, talked things through with his partner, came up with a plan, and quickly made some changes that improved things in the short term. And after that, he was consistently happier in himself—I suspect that exercise had demonstrated to him that he had more agency than he thought and supported him in being more resilient.

The point I want to make here is that you can give people's personal lives space without meddling in them. I did not tell this person how to live their life. Instead, I reflected the impact I was seeing it have on him and gave him space to change it.

I've been fortunate to have a manager who will do that for me. I was grieving after a good friend passed away, and he gave me permission to do the thing I needed but couldn't allow myself to do: take a longer break. He was so right, and I was so grateful. Everyone needs this sometimes, and just because someone is good at noticing this and helping other people doesn't mean they are good at doing it for themselves.

Reduce required emotional energy

When people are struggling personally, the big hit is often to their emotional energy. Maybe normally they're great at juggling multiple things or managing a lot of tasks, but when facing a personal struggle, those abilities are depleted and they're more easily overwhelmed. They may also be more easily stressed by conflict.

Most people are aware that a personal struggle can affect their work, and then they worry about that, which is an additional hit to their emotional energy.

I'm not suggesting that you should give people less impactful work. I'm suggesting you notice and manage the controlling weakness of low emotional energy.[11]

For example, one technique I use when people are overwhelmed in this way is to do a little check-in 1:1 at the start of every day (either sync or async, depending on what works best for them). We go through the things they think they need to do that day and prioritize them. The goal is to figure out what is realistic and identify places where I can help them. Instead of being *nice* and lowering my expectations, I want to be *kind* by helping them focus on the most important and impactful work and support them moving forward every day (even if it's not as much as they would like or think they "should" be).

Even under capitalism, life happens, and almost everyone needs some time off for big life events—especially the hard ones. When people need time away for whatever reason, you can dramatically reduce the impact by making the off- and on-ramps easier. Instead of letting them drift out and drift back, I want to

---

11 We talked about controlling weaknesses in Section 1, "DRIing Your Career"; it's a concept from *First, Break All the Rules* by Marcus Buckingham. Controlling weaknesses are the weaknesses that hold you back, and there's value in getting good enough at them so that they're no longer a problem.

be intentional about maximizing their effectiveness when they are at work and minimizing their worry for when they are not.

Let people be

You cannot manage someone's emotions. You can only give them space to have them—or not. You can only let them know that it's OK—or not.

You cannot fix people, and you cannot fix the situations that happen to them. Any discomfort about that is *yours*; do not fill the space with it. Take it away and feel however you feel about it, but when someone emotes *out* to you, the *worst* thing you can do is give all your emotions about their feelings back to them.[12]

Things to say in the moment include:

- "It's OK; take the time you need."

- "I'm so sorry about this situation. I wish I could do something about it, but I know I can't. What I *can* do is make work as easy as possible for you during this time. I'm going to figure out what I can offer you, and then we're going to work out a plan together."

Agree on the plan together

Consent is important. In most things, it's probably the case that you are the manager and they are "doing the work." You get to approve, redirect, or decide what they do, as appropriate.

However, when it comes to managing someone through a tough situation (whether internally or personally), you need to invert that model. It's your job to do the work, come back to them with options, and let them decide on the plan of action.

For example:

- Say you have a trans employee being deadnamed on some system in the company. It's your job (and, if necessary, your manager's job) to make HR fix it. Do not put it on the person to deal with; no one needs a second job of begging-to-be-treated-with-basic-human-decency-by-people-with-power-over-you at work.

---

12 Psychologist Susan Silk coined the term "the ring theory of kvetching" in her *LA Times* article "How Not to Say the Wrong Thing" (*https://oreil.ly/9v-BO*). There's more about why we do it in Oliver Burkeman's *Guardian* column on the topic (*https://oreil.ly/aPbEn*).

- Medical or parental leave? It's similar. If you're at a smaller company where these things haven't been codified yet, good news—now is the time for you to get them codified.

Your role is to understand the issue, educating yourself as necessary.[13] Then, outline what you think is possible, what you need (in terms of documentation), and what you're going to communicate to the person involved, making sure they're OK with the plan, and go forth and advocate.

If you don't take the time, the time takes you

I had a period in my life when I was in and out of the hospital a lot. I had fractured my shoulder, and some freak optician scan resulted in my having an MRI performed on my optic nerves (everything was fine, except for having an MRI on my head, which was horrible).

After an embarrassing amount of time into this experience, I instituted a rule for myself: days when I was in hospital, I was not also going to work. Didn't matter if it was "just" half a day or some technicality about outpatient appointments. A trip to the hospital is a sick day. End of story.

Having learned this lesson slowly, thinking that I was somehow exceptionally dense and/or lacking in self-esteem, I expected other people to be better at learning the same lesson. This turned out not to be true. The number of times someone has mentioned that they were going to be out for half a day or so for a hospital appointment but would "make it up" in the evening is really extraordinary.

There may be jobs where you can meaningfully contribute when you show up dazed and headachy after having an MRI on your head, but I would posit those are pretty rare in software. If a developer is in a state such that they're more likely to write a bug than fix a bug, they're probably better off taking a nap or watching a movie and coming back rested—and effective. If a manager is more likely to create a problem than solve a problem, to miss something than connect the dots, they too are probably better elsewhere.

In the story that started this section, we saw that though there was an expected amount of time that person might have needed to take off, no one shared this insight with him, *and then that time was taken anyway.* Flexible (unlimited or, more realistically, untracked) time off has some well-documented

---

13 If you don't fully understand something, go and look for resources on the topic. Don't ask for a 101 explainer.

downsides, but if you have it, it allows you to push people to take the time they need to help them be more effective (or just happier) without it cutting into some finite number of days that should be their decision to take.

Drawing the line

Every organization has different standards around performance and handling it, and this includes navigating people going through a difficult time. Some will be more strict, and in others, you will have more flexibility. It will also depend on how valuable that particular employee is seen to be. The level of input or help you can expect from your manager or HR will also be different. Ultimately, you'll need to have your own point of view, one that balances what is right for the person with what is right for the team and what is permissible by the organization.

I want to stress here that these situations are among the hardest you will encounter as a manager. Even the "best" situation—where your manager is helpful, where HR is good—is still extremely challenging. It's very hard to know when to escalate when you're hoping things will improve, and you have a duty of care for the person who is obviously going through a challenging time. However, at some point you have to recognize things are not improving and that inaction is a decision to let them remain as they are.

Life happens, right? And you want to be supportive. But at some point, you will need to evaluate whether you can keep accommodating someone you can't rely on to do the job they are paid to do. Sometimes, a generous severance package is the best and fairest outcome. This is a place where, if you have a good HR person, they are invaluable. A good, experienced HR person will have more context for these situations *in general* and will have a bit more distance than you from *this person*, making it easier for them to be more objective.

If you don't have that, it will be very hard. Here are five core questions to ask:

- What parameters is the organization willing to support?
- When can you realistically expect it to change?
- Is this just affecting the person? Or is it impacting the team?
- Is it undermining the standards of your organization?
- What choices is the person making? Are they making things better or worse?

What parameters is the organization willing to support?

You may have some heuristics about what's allowable, like how much medical leave someone can take or what percentage of fractional time is a minimum. Maybe these heuristics are set in stone or maybe there's room to advocate, but you'll need to decide how much you're willing to.

When can you realistically expect it to change?

Some employee challenges have expected end dates, and you can decide whether the situation can go on for that long. For example, if someone is distracted by moving house, that must eventually end. Maybe the duration of distraction is in the realm of not-great-but-OK. However, other challenges do not have expected end dates and go on indefinitely...which can make it harder to decide when to draw the line. One question to ask yourself is if you knew this situation wasn't going to change, how much longer would you tolerate it?[14] Because at some point, you have to act, and given how long these things take to resolve, you're likely to regret not taking action sooner.

Is this just affecting the person? Or is it impacting the team?

At the point where there is impact on the wider team, the problem has escalated. When you have an IC who is in and out, that affects the projects they are working on, but you may have some leeway there to isolate the impact. However, when you have a manager who is in and out, that affects everyone who reports to them and everything they are responsible for. When their directs are stuck or need something, you're much more likely to have to step in yourself—and now you have to do their job as well as your own. Either way, when there's someone who can't be relied on, that creates resentment from those who *are* reliable and starts to normalize lack of reliability on the team, which can be corrosive.

Is it undermining the standards of your organization?

People look to others around them to determine what is acceptable and what is not. For instance, if you have a staff engineer who is not operating like a staff engineer, then this will cause people to question what a staff engineer is in the organization. The ones who are truly doing the work may resent the inequity there, and the people gunning for promotion

---

14 This question is inspired by the advice columnist Captain Awkward (*https://captainawkward.com*), aka the Marie Kondo of breakups, but I think it can be used more generally.

will compare themselves (favorably) to the person. This undermines your leveling system and generates a degree of resentment that—even in the best case—will be very hard for the staff engineer in question to come back from.

What choices is the person making? Are they making things better or worse?

If someone is making choices that keep themselves in a bad situation, support is no longer support—it's enablement. Your responsibility ends where their choices begin, so if you try to give them options that will help them and they refuse to take them, or they need external help and refuse to seek it out...at some point there is nothing you can do, and you're wasting your time pretending otherwise.

Depending on the answers to this question, you may find you've left the realm of "supporting" and are shifting into performance management. More on that next.

PERFORMANCE MANAGEMENT

Performance management is the umbrella term for communicating with people that they are not meeting expectations, working with them to meet those expectations, and, if they can't meet them, letting them go.

In a similar genre, we might also talk about "managing out," which typically means working with someone to help them understand that they would be better off moving on, without forcing that directly through termination of employment.

Performance management can be difficult, but the reality is that it's part of the job. Avoiding it is a mistake because these things rarely resolve themselves, and they don't get easier with time. New managers in particular won't know how to do it effectively; this is the kind of thing that you are (hopefully) not doing every week or month, and as a result, it will take a lot of time to build up the experience. You may expect HR to handle it, or at least be helpful. While this may be the case in larger organizations, in smaller organizations that haven't yet fully built out the HR function, it may not be, and you will be more on your own. It also differs greatly between organizations. Some organizations aggressively enforce high standards of performance and will default to exiting people and offer exit packages to avoid performance improvement plans (PIPs); Netflix and Amazon are well-known examples of this. Others are slower and more careful about it and default to running PIPs and even moving people around. Within organizations of any size, there's rarely consistency between managers and departments.

I once had a conversation with a new manager who was uncomfortable with the new responsibilities and power of his role. He talked about hoping he would never have to fire someone. I told him it was important to accept that firing people is part of your job. Accepting that is what makes you do the work so that if it happens, you can sleep at night.

The thing about people who hope they never have to fire people is that this mindset makes them liable to bury their head in the sand and ignore the warning signs because they're afraid if they looked at the issues, then they would have to deal with them. Ironically, it makes them more liable to pin the blame on the person who is underperforming rather than taking responsibility for setting and enforcing clear standards for performance across the team—or admitting to making a bad hiring decision.

Of course, unless you're a sociopath, you probably don't enjoy firing people. Firing people is typically the worst-case outcome of a *lot* of work that falls under the umbrella of "performance management." In this section, we'll cover things to look out for and tactics that can be useful. Ultimately, every situation is different. You will try things until something works, or you will try enough things that you're confident nothing will.

Determining ineffectiveness

How do you know if someone's being ineffective? Normally, this is a mix of feedback and investigation.

Feedback is both what people tell you and what they don't.

You need to dig into what people tell you. Sometimes when people give feedback, they have considered their own part in things, and sometimes they have not. For example, if a product manager gives you feedback that an engineer is not getting enough done, you might first want to check that there's clarity around the work and the priorities as well as an accurate understanding of task complexity. You don't want to uncritically relay that feedback only to find that the problem is in sprint planning or technical debt—that will break trust with the engineer.

You also need to notice what people *don't* tell you. For instance, if you have a senior engineer on the team whom you expect to be a resource for their teammates, but no one ever mentions their being helpful, then that may be a sign that you need to look more closely at their behavior. Ask the senior engineer what they're doing, and then ask the other engineers who helps when they have questions or get stuck. It can be really hard to notice when someone isn't doing what's expected, and I think this is a common theme of feedback that becomes a surprise at performance review time. The manager did the work to pass on the explicit feedback, but when it comes time to make a case for the person's promotion or when people are asked for more granular feedback, some things become more visible, such as who has actually been acting as a resource for the team and who hasn't.

By the point when you have feedback from the team (or you've noticed something is awry), you've done some quiet validation and believe it's worth looking into. You may check in to see if anything is going on with them. If they seem to be doing OK as a person, it's time to investigate.

The most useful tool I've found for managing ICs who aren't meeting expectations is to build a timeline. Week by week, identify what they proposed to do (reference your planning materials) and what they actually did (see source control or internal documentation) as well as any mitigating factors (like being sick).

When you plot this out, you can often see patterns you missed. When I've seen someone be really ineffective, often what I see is a series of updates that seem reasonable on a micro scale: underestimating something, getting pulled into something else, or being out of the office. But on a macro level, it becomes a pattern, and it means they are accomplishing very little. If someone was not onboarding well, this helps you map the progress to the milestones and see where things started to go wrong. Maybe they did OK at the starter task, the first easy project, but flailed as they got to the level of work that's expected from their seniority.

The timeline is a powerful tool in situations where you're trying to coach someone or give them more direct feedback and they're telling you that everything is fine (or that the problem is someone else). People are the heroes of their own stories, and they will often put blame elsewhere, which is when having a document that outlines the gap between what they committed to, what they did, and what they communicated is critical for being clear. You need to be in a shared reality to effectively help someone in this situation, and the first step of that is having clarity on what's been happening and why it is a problem.

Building this timeline is also helpful for your communication with other people, like your boss and HR. It helps you get on the same page about what's actually been getting done and how that's being communicated. If you don't have the information to put together a timeline, that would be a helpful thing to start putting in place before you need it. There'll be more on building practices around planning and communication in Section 4, "Self-Improving Teams".

Identifying toxic behaviors

Think about the people in your career (or outside of it) whom you have encountered that you consider toxic. What do they have in common?

My working definition is that toxic people are trying to preserve their own status at the expense of everyone else. Sometimes, this looks like arrogance: thinking they have all the answers and thus devaluing the contributions of other people. At other times, this looks like trying to tear things down in order to keep things the way they are. When they give feedback, they often make the person they're giving feedback to feel smaller. When they complain, it often slows or prevents change. Their questions, especially in public settings, are more about sowing dissent than the answers they receive.

It can be very hard to deal with those people. They create paper cuts that divide the team in small ways. Complaints about any specific instance of behavior feel petty; the instance itself is unimportant and a waste of time. However, the overall impact can be significant. It drags people down, and if you're trying to create culture change (and culture change is an invariable requirement for scaling), it can feel like there's a large rock that you have to work around and drag along.

In a distributed environment, these people can make themselves feel bigger than they are: sowing dissent via backchannels, in anonymous forums, such that you don't really know where it's coming from or how prevalent it is. Is it one bitter person who shouldn't be listened to? Or is it multiple people, and there's something to change and/or clarify?

In *Trust Your Canary: Every Leader's Guide to Taming Workplace Incivility* (Bar-David Consulting), Sharone Bar-David defines *workplace incivility* as "seemingly insignificant behavior that is rude, discourteous, insensitive, or disrespectful, with ambiguous or unclear intent to harm the target." The book quantifies the impact of this, which is higher than most managers think, and there's a list of behaviors, including:

- No hellos or skipping greetings
- Belittling of opinions, experience, and skills
- Talking down
- Sarcasm
- Cliques, gossip, social exclusion, or shutting someone out
- Dismissive, excluding, or judgmental body language or sounds

- Passing blame for mistakes onto others
- Stealing ideas and taking credit
- Withholding information
- Asking for input and then ignoring it

Ultimately, this kind of behavior can be very hard to address, especially in environments where it is also exhibited by leadership. Understanding the behaviors and documenting them, even if just for yourself, can stop you second-guessing the extent or impact of them.

In a distributed environment, consider ways to prevent people from making it seem like their attitudes are more prevalent than they are. For instance, run a survey around team health metrics so that you can understand the extent of the problems.

The PIP

If you have an HR department, they will likely have a process for running a performance improvement plan (PIP), which is a process to document and address poor performance. Remember that a PIP is fundamentally an adversarial process where both the manager and the employee have to prove that they are doing their jobs. HR will not take this off your hands; they will put you in a process where you are forced to be very clear about what you expect someone to do and deliver regular, granular, corrective feedback. Support from your manager is often critical in managing this, and if you are managing a manager who is running a PIP, make sure to be there for them.

The first time I managed a manager who had to deal with a PIP, the process was so hard on them because of the volume of work it entailed and the lack of clarity from HR that they ended up stepping down from a leadership role, and as time passed, I saw this happen in other parts of the organization. I learned an important lesson and made sure that subsequently, I intervened more, ensured that the manager got the support they needed and that the desired outcome was agreed on with HR (i.e., the person working independently without need for micromanagement), better coached the manager on how to handle the situation, and invoked my boss as needed.

Here are some things to keep in mind when navigating a PIP:

- Communicate upward to your manager about what is happening, what you're doing, and why. Ask for their advice and support: first, because you want them to hear it from you; and second, when you give someone performance feedback, their first step may be to go to your boss and complain, and you will want your manager to reinforce your message, which means they will need to know (and agree on) what that message is. Your manager can also help you understand how and when to involve HR.

- Start a timeline early. Don't wait to put this together; from the moment you wonder if it might be necessary, start writing things down as they happen. Having clarity about what is going on and for how long can make assessment of the extent of the problem clearer.

- Start documenting your feedback to the person in writing as soon as you realize a PIP is a possible outcome. This will help if you're inclined to soften your feedback verbally, and following up on the conversation in writing will make the severity and action items clear.

- Make sure that when you set outcomes for the person, they are the actual outcomes you want to see. I once saw someone PIP'ed on their communication, which got them performing better at communicating...but the fundamental problems of how they approached their job remained.

It's easy to write something clear about the mechanics of what the employee needs to do, but the outcomes are more important. If the PIP simply says, "Write a status update each day," and the person does that, they can be declared to have passed the PIP, even if they are still not delivering what you expect them to. Spend the time on writing the PIP to focus on outcomes, push back on HR if you need to, and invoke your boss if necessary. However difficult this is, it's easier to handle this before delivering the PIP than after.

The PIP can end in different ways:

- During the course of the PIP, it can become clear that the person is not meeting the expectations, and so they are let go.

- At the end of the PIP, you can conclude that the person is still not meeting expectations, and either (1) you decide whether you want to let them go, or (2) if they're making progress and you want to invest more time in them, you might extend the PIP period.

- The person can get back on track, thus meeting the conditions of the PIP.

During a PIP, you will feel like you are micromanaging someone—by definition, you are. This is a *temporary* state, where you have to prove you are being extremely clear about expectations, and the person has to prove they are meeting them. When the PIP ends, the expectations remain in place; if someone later breaks them, then typically you would not need to run another PIP in order to terminate their employment.

Rehabilitation

People do change; it's just not always within the time frame that you want them to. The biggest mistake that managers make when it comes to addressing people's effectiveness (or behaviors) is that they wait until they have given up on the person before they try to address what they want them to change.

What has helped me is to think less about this in terms of "escalation" and more "What will it take for this person to hear this?" If the coaching approach hasn't worked, then it's time to try something else, such as the authoritative and eventually the coercive style (we discussed how to expand your leadership range in Chapter 8). It can also work in the other direction: sometimes the authoritative style doesn't work and what does work is a softer approach, such as coaching them (or sending them to a coach) into more self-awareness.

It also helps to remember that someone's first reaction is not necessarily their final reaction, which is a lesson from *Thanks for the Feedback*: the first score (how they take the feedback) is distinct from the second score (what they do with the feedback).[15] Once I experienced someone scoring a zero in taking the feedback and then a 10 in what they did with it, my relationship with hard conversations was changed forever.

Finding Balance

When we talk about "attrition," we tend to talk about "regretted attrition" and "unregretted attrition." These are polite phrases that mean people we wanted to stay and people we were happy to see go.

Normally, this wording of "unregretted" masks so many different aspects. It covers the people we fire and the people we "manage out" as well as the people who weren't doing either of those things but then, once they were gone, everyone was *relieved*. And often it includes the people who were just...*fine*. Not *great*, not *terrible*, just...*fine*.

---

15 Again, this is a great book that is widely applicable!

So many questions:

- *Why* were they ineffective?
- *Why* were they toxic, and how did it get to that point?
- *Why* were they just...fine?
- *What* competencies were missing?
- *What* happened in the hiring process?
- ...*What* changed?

We often talk about these things like they are about individuals, but the environmental factors are very strong. In watching multiple team transformations, I've noticed that what shifts is how performance is perceived. The individuals were the same, but as the environment became more functional, they in turn became more effective. Their roles hadn't changed, but how they were positioned and how effective they were within that environment did change. At the other end of the spectrum, other people were too reliant on the old environment, couldn't adapt, and eventually moved on. Unregretted.

Given some line of expected performance, some of your team will be above it, some around it, and some below. You need to figure out how to apportion your time to these groups effectively and use that time to move people up. If you give all your time to the people below the line, you are neglecting everyone else (and likely making yourself miserable). If you give all your time to the people above the line, then you're not addressing problems that exacerbate inequity, and you're underinvesting in people who could be above the line with more direction. Effective, hard-working people tend to get pretty grumpy about ineffective people, so you'll be damaging team morale, too. Making people successful—from both inadequate to up or out and good to great—is a core skill required to have an effective team.

Although letting someone go can be sad and hard, and may provoke some upset on the team, it can also be a huge relief to no longer be dancing around the difficult person or no longer pretending that you expect the unreliable person to do the work they committed to. It's probably not great for someone to be in an environment where, on some level, they know they can't succeed, but you also don't need to tell yourself some story about how you "fired them for their own good"—it's your responsibility to make the decisions that are right for the team. Your responsibility to the individual was to give them clarity about the situation and the opportunity to make different choices, but the choices they made were their responsibility.

Being good at leadership is not about everyone liking you all the time. Sometimes you need to protect someone so that they can have some space to recover. Sometimes you will need to say the kind thing rather than letting the person off with *nice*. Sometimes you need to challenge people who do not want to be challenged. Sometimes you need to set ever-clearer expectations and find it in yourself to believe that someone can do better. And sometimes you need to let people go.

People do not grow in a consistently linear fashion. They have different inclines, depending on a multitude of factors. Sometimes they are all in on the stretch assignment, pushing through a high-growth period. And sometimes they are flatlining, or even contracting, to create space for other things in their lives.

Being an effective manager of a human being means trying to understand where they are at and what they need to be successful. It means trying things, and sometimes failing and iterating, until you figure out what works. It also means knowing when to cut off that level of effort and start documenting things. It is more art than science, and everyone gets things wrong. It can be a hard skill to develop because the tools that work with one person may not work with the next one. Maybe nothing would, but maybe you just didn't have the right range. You will never know for sure, so focus on what you can learn from the situation and apply it going forward.

Building a Bench

As teams grow, they hit different bottlenecks and limitations. The DevEx requirements of a hundred people are fundamentally different from the requirements
for a few people, and those of a thousand are different all over again.

Similarly, the leadership requirements change as organizations scale. Small
organizations may "not need managers" and have some flat structure where
every engineer reports to the CTO and think it works. Invariably and inevitably,
this structure will break down. It will break down for the first obvious reason of
just how many direct reports one person can have, and it will break down for the
less obvious reason that everyone, while in small, highly coupled environments,
can know what is going on, at some point you cannot—and do not want to—
onboard new people into a situation where in order to be effective, they need to
know everyone and everything that is going on. And finally, it will break down
because the profiles of people you will hire will necessarily need to change. Early-
stage startups may value flexibility and scrappiness over skill and experience,
but at some point, you will need to hire different profiles. You will need some
different skills to scale backend services or to turn the basic app that validated
product-market fit into a polished experience. Different people will want to join a
team with 6–10 members than the people who want to be the first engineer (of
that function, or period). Both types of people will have different expectations.

Early-stage companies, and later-stage companies that have decided to rid
themselves of org chart layers and meetings in response to shareholder activism,
may describe managers as "overhead." It's easy to resist this labeling, but I
believe that we are better off accepting it. Engineering managers are overhead,
in the sense that they are an extra person who does not obviously contribute to
the bottom line by shipping features and fixing bugs. That means that we need
to make that overhead worthwhile. That is, we need to ensure that it is clear that
our teams deliver more *with* us than they did *without* us, that things are more

organized, that it's easier to understand the status of things, that there are vastly fewer (ideally, no) instances of some developer going off down a multimonth rabbit hole chasing a feature that no one really cared that much about anyway.

This need to demonstrate value often means that when we add leadership layers, we do so too late.

There are some heuristics in management, like numbers at which teams should split or the maximum number of people that someone can realistically manage. These are contextual and depend on the organizational structure, the complexity of the work, and the competence of both the people and the peers.

Maybe it's possible to manage 10 developers senior enough that they need minimal direction when you have a very competent tech lead (TL) and product manager (PM) and managing that team is your primary responsibility. It won't take much change for this situation to become impossible: the PM quits or the TL is needed elsewhere, or you swap two seniors for two juniors, and the job that was (just) manageable will go (rapidly) to hell.

This is why we need to build a leadership bench: people who can take on more responsibility as the team grows. A bench gives us *slack*, and that slack is the shock absorption for the unexpected events that happen.[1] If we think about who *can* step up before we *really need* someone to step up, we will have that much more time to help them grow into it.

As teams grow, we will need more managers, and we will need to figure out who those managers might be...and we have to identify them well in advance of whatever concrete opportunities arise. Before that, we'll need people who can mentor, who can onboard, who can participate in hiring. We'll also need people to step up and lead *technically*, taking on increasingly large projects and stepping into a TL role. There will be some overlap here; your best onboarding buddy may also be your best technical lead, and that makes sense—there are overlapping skills there. However, you won't want to drop everything on one person; you'll need to figure out a mix of people to whom you can delegate different things. To make it harder, you won't know exactly how things are going to play out, so you won't be able to promise *outcomes*; you'll be creating the kind of *optionality* we talked about in Section 1, "DRIing Your Career" for the team and the people on it.

---

1 More about this concept in the book *Slack: Getting Past Burnout, Busywork, and the Myth of Total Efficiency* by Tom DeMarco (Crown).

This is also the manager's perspective of what we talked about in Chapter 1, and in Section 1, "DRIing Your Career" more broadly. Individuals (typically) need and want opportunities to grow; otherwise, they will look elsewhere. Leaders, particularly of growing teams, need and want people to take on additional responsibility. Interests are aligned here, and yet it doesn't always work out. Why?

So many reasons, such as:

- Waiting too long to delegate
- Failing to provide sufficient direction and support
- Neglecting to develop a strategy
- Developing a strategy that is overreliant on one person
- Underestimating the overhead generated by complexity
- Resisting giving up control

As we talked about in Chapter 6, having a strategy is part of the job. If the team is scaling or you want to be in a position to take a different role should the opportunity come along, you will need to scale yourself, by either growing people internally or hiring them in. So in this chapter, we'll cover some critical components of that: how to identify and develop potential and when and how to hire in leadership.

Identifying and Developing Potential

Hiring external folks into leadership roles is a *lot* of work, and it takes time. Typically, it's also something you can only justify *after* it's necessary, which can make it hard to get ahead of it, and then add onboarding, which also takes a long time. This is the practical reason to look first at the team you have and identify who you can develop; these are the people already there, no waiting required, and they already understand how the organization works. Then there's the longer-term, downstream reason: if you only hire in leadership, you send a message to the team you have about their potential for growth. Finally, the values reason: developing people is part of leadership, and it behooves you to make meaningful effort there.

If you come in with minimal context, figuring out who are the people you can develop quickly can make a huge difference in your ability to get things in order and be effective and can also be a shortcut to building buy-in with the team. In struggling teams or rapidly scaling organizations, the lack of individual development can be glaring, and it can be a meaningful win for demonstrating how things will change under your leadership, as people see more immediately how the change will benefit them personally. So in this section, we'll talk about how to identify potential and how to develop it once you find it.

The Role of Belonging in Growth

On top of the belonging and accomplishment we talked about in "Onboarding" on page 212 sits growth. The belonging and accomplishment are like a survival baseline, but what goes on top of that is much more interesting.

We've all heard the saying "People don't leave jobs, they leave managers." Actually, research from the University of Chicago (*https://oreil.ly/LK1Br*) shows that people leave good managers and bad managers at about the same rate. The difference is largely that when people leave a bad manager, they leave *because* they have a bad manager. When people leave a good manager, they go on to a better opportunity: an opportunity that the time with that good manager has set them up for.

Either way, people leave because they want change. Sometimes that change is about the baseline; they want to escape an ineffective or even an abusive manager. Sometimes they leave because they want their life to be different in a way their current role doesn't fit with—something we saw a lot as some companies embraced remote cultures while others tried to bring people back to the office. And sometimes they leave for a bigger opportunity than you can give them.

Fostering a sense of belonging and accomplishment helps retain people. Why?

- You don't just do belonging for one person; there has to be something coherent to "belong" to: the team.

- Onboarding, as we discussed in Chapter 9, through the combined lenses of belonging and accomplishment, shapes team culture, creating a better place to work.

- When people belong, they can fail—and learn (more).

- When you democratize accomplishment, you remove the zero-sum competition. People are invested in one another's success rather than competing with one another.

Some career advice I got about a job was that it would be "not fun in a shrinking market." It's the corollary of the "get on a rocket ship" advice: growth creates opportunity while shrinking markets remove it, creating more competition, which results in fewer team-minded behaviors. I joined the company anyway, but when I started thinking about leaving that job, I came back to this advice; it was prescient. Not immediately true but eventually true.[2]

I always felt that the rocket-ship advice was missing something, though. Here's what I think it is: while growth creates opportunities, belonging supports people through them. Whether the rocket ship is exhilarating or exhausting depends on that sense of belonging. This is why strong teams—with that sense of belonging—can do truly amazing things. And this is why many employees struggle at companies in high growth: the companies that haven't nailed the basics of belonging or where the organization's growth has undermined the sense of belonging. This is why onboarding is so critical: because the time to start with belonging is on the person's day one. It gets harder and harder to address over time. Starting on day one will mean they are in a position to take on more, more quickly, and will be more inclined to do so.

IDENTIFYING POTENTIAL

One definition of *potential* is the gap between what someone is doing today and what they are capable of.

How do you find it? By understanding people's strengths, weaknesses, and coachability. I like to use the following questions to find untapped potential on a team. For each person, ask (yourself or their lead or manager):

- What are they great at?
- What do they help other people with?

---

2 I guess good advice is eventually consistent rather than real time.

- What do they need help with?
- What's their potential?

A core part of this is that you will need to notice these things, and not all of them will surface in explicit feedback, so you need to pay attention to the implicit feedback and, ideally, build a culture of feedback that makes these things more visible.

Like so many things, identifying potential is fraught with bias. It's easy for the "high potential" person to be the one who looks like you, who talks the loudest, who self-promotes the most. If you're not mindful about this, you set up a bad system of incentives and leave much potential untapped.

FEEDBACK CULTURE

The biggest issue with trying to create a "culture of feedback" on a team is that people often see "feedback" as synonymous with "criticism." As such, one of the most effective ways to build a feedback culture can be to neutralize that concept and instill a new one. Feedback is just how you show up in the world reflected back to you—this is why so much feedback ends up being about communication, because communication is so much of how we show up in the world and, as such, has the largest surface area. To help with this, try focusing on three Cs of starting to build a feedback culture on an engineering team: code review, calibration, and compliments.

Code review

In software, we have a huge head start on creating a healthy feedback culture, in that code review is a normal part of the job. As a result, most developers quickly learn to separate themselves from their work (or at least their code). When they work at being better at code review, they also work on their feedback skills generally, because those skills map. For instance, giving a structural pass before a detailed pass in order to be more effective with their time. Or creating documentation or norms (code style! linting!) such that a machine can pass on the team agreements, rather than a human.

The challenge is that people often leave the code review and go to some other situation without realizing that all those skills map. But that is easily remedied through awareness and coaching. Rather than treating some other situation as a new skill, map what they do well elsewhere: "I've noticed when you do code review, you always do a structural pass and then a detailed pass—why not try that in this situation too?" "I've noticed when you do code review, you clarify what comments you feel strongly about and what you don't. Can you do that here as well?"

On a team level, developers can usually be sold that improvements to code review will positively impact team effectiveness. That can make it a great place to start. And if there are problems on the team, they will usually show up in code review, making it an incredibly helpful place to get a read on and start coaching a new team.

Calibration

In Chapter 8, we talked about the importance of calibration—ensuring that people understand what you're looking for and that they are consistent and fair in applying it—in an effective hiring process. The other benefit is that you have a tangible, close-to-real-time way to coach someone on articulating feedback. When you debrief on an interview or the review of a code test, it's an opportunity to ask questions that help someone clarify their observations and impact.

By digging into what someone has noticed and helping them articulate it clearly, you can move from something vague, like "I don't think they would be successful on the team," to something more concrete, like "They seem to have worked mostly alone on less complex applications. They didn't seem very curious or aware of the gap technically, which makes me concerned they don't appreciate the learning curve they would have and seem disinterested in what it means to work on a team and the team we have here. If we decide to move forward, we should dig into those aspects and better ascertain their willingness and ability to adapt."

The thing I love about using hiring as a mechanism for helping people improve with this is that it's a very efficient way to help someone work on the skill of articulating feedback: one 60- to 90-minute interview, about 15 minutes to debrief, a few small suggestions...and then come back a little later to see how they're doing and give them some more suggestions—or compliments (or both!)

This also applies to any feedback. It's annoying to get vague feedback at review time, but you can also use that to coach people on giving better feedback. Giving feedback, like any skill, requires time, attention, and practice to improve. The more access you get to someone's feedback, the more you can help them get better at it and encourage them—where it makes sense—to deliver it themselves.

Compliments

I firmly believe that a healthy feedback culture is built on a culture of appreciation, which is where *compliments* come in, aka "positive feedback" or "reinforcing feedback." The value of this is that people are more open to thinking about how they can improve when they feel like what they *do* do well is appreciated. Also it makes for a nicer place to work.

Model this behavior by giving people positive feedback *at the same level of specificity* that you would give critical feedback. "Great job" is woefully insufficient. Try: "You did a great job presenting the architecture proposal to the team. I really appreciated how you contextualized the problem in order to start on the same page. You outlined the trade-offs thoroughly and made the rationale for your recommendation clear. This really helped keep the discussion productive. Thank you!"

Encourage this behavior by instilling four words into everyone's head: "Have you told them?" If you say this consistently every time someone says something nice about someone else, people will start saying it to one another—and themselves. Building out systems of positive recognition (like the "snaps" posts mentioned in Chapter 9) is another way to encourage this mindset.

Changing culture

Building a strong feedback culture takes time and reinforcement, but these three levers will give you a good starting point. They cover a lot of the same skills as giving constructive feedback but in a way that people are considerably less likely to find threatening, with minimal potential downside (how badly can a work-related compliment backfire, really?).

Once you've made some inroads there—people know how to phrase feedback effectively and give it promptly, and they have built good relationships—you can push for them to give one another more constructive feedback, if you need to. But note that you will have gotten a lot of what you need in terms of identifying potential, having a better read on people's coachability, and understanding the things their teammates appreciate about them.

DEVELOPING POTENTIAL

In Chapter 9, we talked about making the most of your 1:1s. This is foundational. As teams grow, though, you will also need to develop people that you don't have regular 1:1s with. This means that while you will interact with them and offer feedback, you won't have the same depth of relationship and regularity of communication. You will need to set up your rewards and incentives such that they incentivize the behaviors you want to reward—and discourage the ones you don't. You will need to coach their manager to coach them, offering feedback at one level of indirection, and push the right opportunities toward them—especially the ones that their manager may not have the same level of insight into. As you do this, you will need to be careful not to undermine your direct reports; they are closer to what is going on and better understand the day-to-day, so talk to them first and make sure you're aligned. You will also need to evaluate the managers reporting to you on their ability to develop potential effectively (and equitably!), pushing them to build their own bench.

Identifying suitable stretch assignments

In Section 1, "DRIing Your Career", we talked about finding opportunities for growth and the idea that growth was where we left our comfort zone. When developing others, you have to figure out how they want to leave their comfort zone. One way to do this is via the "stretch assignment," the opportunity that will allow someone to take on more than they have done previously, demonstrate their capabilities, and learn.

What makes for a good stretch assignment?

- It's actually a growth opportunity.
- It will be recognized.
- They will be set up for success.

For something to be a growth opportunity, it has to match the person's growth goals and be meaningfully different from things they have done before. Here are some examples of growth opportunities and some that are not:

- Having someone who previously led implementing features own an entire application or service? That is a growth opportunity.
- Asking an IC to take on a technical lead role? Growth opportunity.
- Asking a director to cover the VPE's parental leave? Growth opportunity.
- Asking a director used to managing teams of teams to get hands on and deliver something? That is not a growth opportunity—it's a favor to the organization.[3]
- Asking a tech lead to join another team to be an IC and get something over the line? Also a favor.

---

3 It's fine to ask someone to take one for the team, but if you try and gaslight someone into believing something is a career opportunity when it obviously isn't, they will be justifiably annoyed.

Recognition is personal, and it's important that we understand what it means for someone to *feel* recognized. You will want to make it clear what you can and can't do. Be mindful of someone taking "this will be recognized" to mean "this will get you promoted." We'll talk more about this later in the chapter.

Setting people up for success is probably the thing you have most control over. You can define the problem space, identify support (like a mentor or an external coach), and ask them what they think they need to be successful. Be careful that you don't overpromise here. If you commit to something but don't provide it, the person will feel—understandably—like you hung them out to dry.

Some people thrive on being thrown into the water and left to sink or swim. But that's not appealing to everyone—and it can be particularly unappealing for those for whom failure has a much higher cost.[4] Making the prospect safer makes it seem more possible for a more diverse set of people. Bringing the conversation of support up front makes it less like it's addressing a problem and more like a normal part of taking more on.

Story

Just a year out of college, I had somehow landed myself on a team that was building a zero-to-one product under the guidance of several seasoned industry veterans whom I would later recognize as my earliest sponsors. One of these industry vets was Doug. He was among the most senior engineers at the company and our team's technical lead, an unofficial role that carried a number of loosely defined leadership responsibilities. Two of my other experienced teammates were Alan, an engineer with more years of programming experience than I had years on Earth (literally), and my boss Greg, whom I'd met during my interview and clicked with immediately.

All four of us worked together on this product for several years (of course with the help of many other talented teammates), seeing it through its initial launch and subsequent user growth. The product had become a popular fixture of the company's platform offerings. Now that things had stabilized, Doug, who was great at delegating work to me and had advocated hard for my promotions in the years prior (a sponsor

---

4 Under-indexed people; see also: the glass cliff (*https://oreil.ly/nSSZI*).

before I even knew there was a term for it), was unsurprisingly tapped to lead a new major project on another team. I was naively terrified for us to lose him.

The only thing that put me at ease was the assumption that Alan would step in to take over Doug's technical lead responsibilities. It made sense to me: Alan had the most industry experience on the team, he'd been with us since day one, and he could answer pretty much any question Doug could answer. I looked up to him the same way I looked up to Doug. So imagine my shock when Greg asked me in our 1:1 to take over the technical lead role. Without hesitation, I turned him down. "Why wouldn't you give this role to Alan?" I asked. Greg told me that he and Alan had already discussed it, and not only did a lot of the responsibilities not appeal to Alan, but they both thought I would benefit more from the opportunity. With my impostor syndrome kicking into high gear, I insisted that Alan take the role, stating, "I don't know enough to be a technical lead; I can't do it." A bit insulted, Greg exclaimed, "You think I would offer you something I didn't think you could do? How would it make me look if you failed?" He had a point.

Over the next few days I got comfortable with the idea, with Alan and Greg (and Doug too!) assuring me that they'd be there to support me any time I felt like I was in over my head. I accepted the role, which eventually carved a path for me into management. Looking back on it, I can see how fortunate I was. Not only were they willing to give me an opportunity that could have easily gone to the more experienced team member, but they were adamant in convincing me when I tried to refuse it. A more confident engineer in my same position might have eagerly jumped at the chance. A less thoughtful manager might have accepted my "no" and moved on to Plan B.

I have always carried this lesson with me when sponsoring others. Sometimes I need to offer people an opportunity they wouldn't otherwise ask for or consider themselves capable of. Sometimes I need to assure them of my belief in their abilities and lead with how much support they will have to be successful. And many times I have found myself channeling Greg and dramatically asking why on Earth I would let them fail when that would only reflect poorly on me! That one works every time.

—*Jill Wetzler, VP engineering and consultant*

I've come to observe that sometimes those who feel they need the most do the best over the medium to long term. They are more likely to embrace the help available to them, work harder to overcome natural preferences, and pay that support forward to others on and off their team. It can be a leading indicator of those who will level up and those who will burn out.

You may think you don't have time to help people succeed, but you probably don't have time not to. It's a lot more work to unravel a situation—particularly a leadership situation—that isn't working out than it is to set it up well in the first place. When delegating something, a rule I use is to expect to spend a third to half the time that it would take me to do something badly on helping someone else work up to doing it well. Framing things in that way helps me communicate expectations around what people can expect, getting them to accept the help more readily—it's not failing, it's onboarding! It also helps me be realistic about how much I'm really giving away and on what timeline.

Note

It's worth talking here about how a "just take ownership of what you want to be better and trust it will be recognized" mentality can advantage some people over others. This is because of the "authority gap," which is a measure of how much more seriously men are taken than women and how women are generally punished for typically "masculine" leadership traits;[5] or as Jo Miller, author of *Women of Influence* (McGraw Hill), puts it, "Women face an exasperating double bind: self-advocate, and be sidelined for lacking social skills. Fail to self-advocate, and have your competence questioned." Although these data points focus on women, it's a safe assumption that racial inequity will also show up here.

Developing warning signs and intervention points Stretch assignments should not be set and forget. You want to make sure that you have visibility into how things are going and set up points where you will check in and intervene if necessary. This will include things like:

- Holding regular skip 1:1s for teams with new managers
- Reviewing roadmaps earlier and more actively for new tech leads

---

5 Michelle P. King, author of *The Fix: Overcome the Invisible Barriers That Are Holding Women Back at Work* (Simon & Schuster), writes about the authority gap (*https://oreil.ly/FY_8y*).

- Having additional interactions with people taking on other responsibilities, such as more frequent 1:1s

Recognizing growth Under capitalism, money is the way we communicate value. As such, valuing someone often gets framed as having some kind of dollar value.

Conventional wisdom has been that above a certain baseline, more money does not make people happier.[6] However, it's the most obvious lever and the one that managers are most likely to jump to. Sometimes people just want more money. But sometimes, as we discussed in Section 1, "DRIing Your Career", what people want is more nuanced.

This means that often when it comes down to rewarding, the manager and the employee end up having the wrong conversation. The employee worries they aren't growing, but the conversation becomes about how to get promoted (and "managing expectations"). Or the employee doesn't feel valued, and the conversation becomes about compensation.

Compensation is important, and it is important to pay people equitably. But if that is your primary lever for making people feel valued, you're going to have a bad time.

This is why many companies follow a compensation approach that reduces—or even eliminates—negotiation. Negotiation is generally more risky for under-indexed folk, so it's an equity issue, but it also has the potential to be an ongoing distraction.[7]

Although compensation and promotions are important, they happen infrequently, so unless someone is significantly underleveled or undercompensated, how much people feel valued day to day is typically more driven by their day-to-day experience. This includes:

- How often they receive positive validation, the specificity of that validation, and (for some) how public it is

- How much time you invest in them—how often you reach out and how willing you are to help

---

6 The original study on this has since been followed up with a more nuanced result (*https://oreil.ly/0UxQ2*).

7 There's a lot of great information on this in the book *Women Don't Ask* by Linda Babcock and Sara Laschever (Princeton University Press).

- What you do for them—for instance, finding them additional training and proactively getting it approved by HR rather than waiting for them to ask for it

If you want someone to feel more valued by the organization, you probably need both of the levers of capitalism: compensation and promotion.

If you want someone to feel more valued by *you*, you have many more tools available to you. To start, try telling the people you work with directly something you value about them each week in your 1:1.

If you want people to feel more valued by their teammates, you can encourage practices around positive feedback and peer support that will create a positive culture.

Whenever we talk about rewarding, it's important to note that the behaviors that are rewarded become part of the culture of the team. When people are rewarded for diving saves but not for smooth rollouts, there will be more diving saves. When people are rewarded for shipping shiny features but not for the less glamorous work of bug fixing, there will be a focus on shipping shiny things, and bugs will go unfixed. As such, and we'll talk more about this in Section 4, "Self-Improving Teams", I encourage you to pay attention to and reward the behaviors that support good team functioning. The person who quietly handles many releases. The person who volunteers to be the onboarding buddy for the new hire (and does a great job). The person who pays attention to incoming bugs and addresses them alongside their current project. I'm continually amazed how much people appreciate these things even being noticed at all.

Making problems tractable

The more responsibility you accumulate, the more your role will be around navigating ambiguous problems. Your ability to untangle and resolve ambiguous problems may even be core to what has propelled you up the ladder. As a result, it's easy to think that the way to develop other people is to give *them* ambiguous problems on the basis that they will then develop this skill.

Sometimes this works. Often this doesn't.

At the core of the right level of responsibility, or the right stretch assignment, is someone feeling like the problem is *tractable*. This means they can reason about it and have some ideas of what to try in order to understand it better and/or push to resolution.

In many ways, this is similar to programming. When you get a feature that you don't understand and don't know how to build, you first have to figure out the definition of the feature, what you need to know, how you will validate what you're doing, and so forth. In time, maybe you graduate to building systems, and then maybe you graduate again to resolving interdependent systems. Intuition and experience help you figure out what to investigate, what to document, what to prototype, what discussions to have, and how to resolve them effectively.

To pick a very meta example, say you need to scale a team, which is an ambiguous, complex problem: the hiring process is probably wrong, onboarding is probably insufficient, and org structure probably needs to change, as do the projects and the distribution thereof. This is an open-ended, ambiguous problem that you will need help to solve. However, you don't have a process to hire in leadership, and you don't have anyone currently on the team who can take any of these problems as is.

Enter the move I call "clarify and shard." Your role in each problem is to understand it well enough that you can clarify it and hand it off to someone else.

This might look like:

- Understanding hiring well enough to identify bottlenecks and produce some rough definition of rubrics, then giving it to one or more people who can (1) resolve the bottlenecks and (2) refine the rubrics and calibrate people on them

- Identifying some core issues with onboarding, instituting an onboarding buddy program, and picking and supporting your first onboarding buddies such that they can (1) effectively onboard your first few people and (2) set up some basis for a working system of guidance and expectations so that other people can take that role on

- Implementing a minimal process around project allocation and clarity, and letting others refine and build on it

- Keeping org structure because no one else has the level of breadth to do it effectively (but having got everything else set up to improve over time, this problem is much more tractable *to you*)

When we want to parallelize programs, we need to unravel the dependencies and interconnectedness, so we can give one process A, another process B, a third process C. People do not need quite as much clarity as computers do, but they still need clarity. We will talk more about *creating clarity* in Section 4, "Self-Improving Teams".

REFLECTION

Think of a problem you are currently having to handle yourself. What would you need to do to make it tractable to someone else on the team?

Navigating performance reviews

We can't talk about growth without talking about performance review time—that time of year where everyone is extra stressed and, as managers, we reap the benefits of all the things we did well over the year and pay the debt of all our missteps. Fun!

The first rule of performance reviews is *no surprises*. This is where the work you do the rest of the year really pays off. If you're having ongoing conversations about growth and expectations and giving regular and consistent feedback, then performance review time for most of your directs probably won't be that hard.

I say "most directs" because it is still possible to surprise people if they haven't heard you even when you have told them things. Maybe you were genuinely not clear enough and can learn from that, or maybe they just chose not to hear you.

It's also possible to surprise someone if you are also surprised. For instance, you might find that things show up in 360-degree feedback that haven't previously been mentioned. The less involved you are in the details of someone's work, the more likely this is to happen. At this point, you will need to dig into the feedback you're getting. Is it fair? Why hasn't it come up before? You will need to decide what to do with it during this review cycle, including how to relay it to the person and how much weight is fair to give it in their review. You will also need to figure out how to prevent the surprise feedback next time—whether this involves giving explicit feedback to the person who didn't speak up, or taking it as implicit feedback for you about checking in on certain situations in more depth, or some combination of the two.

The second rule of performance reviews is *everyone gets to feel how they feel*. Evaluative feedback that directly affects compensation is hard, and the prospect can be difficult and scary—especially when people have had previous bad experiences with such things. It's worth noting that you, the manager, also get to feel how you feel.[8] This can be a really hard process, and giving yourself space for

---

8 But be *very* judicious about who you share those feelings with!

your own feelings such that you can set them aside when you're with your directs is key.

The third rule of performance reviews is that *first reactions are not final reactions*. In *Thanks for the Feedback*, Douglas Stone and Sheila Heen talk about the two scores of feedback: the first is the feedback, and the second is the response to it (we touched on this earlier in Chapter 3). Even when people respond badly in the moment, they can react really well to the feedback over time. Equally, some people can seem to react well in the moment but in a way where they don't ask the questions they need to, leaving them unresolved, which results in their moving forward without the clarity they really need.

Good feedback should set people up for the next cycle. Let's talk about three kinds of cycles:

- Lay the groundwork
- Close the gap
- Build your strengths and thrive

Lay-the-groundwork cycles are foundational. For example, any onboarding cycle is typically a groundwork cycle. Or a cycle after a role switch, such as into leadership. After org changes, people may need to lay the groundwork again (this is part of why reorgs can be so disruptive to people's careers). This is about nailing the fundamentals of the job and setting the person up for long-term success—ideally going into a "thrive" cycle.

When running a groundwork cycle:

- Make sure the person has time and space to focus on the fundamentals of their job.
- Make sure they get corrective feedback as necessary—the earlier the better.
- Pay attention to their confidence level, ensuring that they are building confidence over time (and are not overconfident).

Close-the-gap cycles are about addressing some—ideally, well-defined—deficit. The key thing is some clear and understandable definition of what it takes for a person to close the gap between where they are now and the next level (or, if it's minor and doesn't merit a full performance management setup, their current level). If that can't be defined, it's not a close-the-gap cycle. The *worst* experience for someone is to close the gap they think has been agreed upon only to learn of another gap next cycle—this makes people feel like the goalposts are continually moving and question if they will ever be successful.

When running a close-the gap-cycle:

- Make sure it's really clear.
- Make sure it's accurate (no moving goalposts!).
- Pay attention to it throughout the cycle, making sure that progress is demonstrated and clear and that the required people recognize it.

Build-your-strengths-and-thrive cycles are about leaning into strengths and building on them. They are the most fun cycle! The key thing is that someone needs to be operating from a position of strength. If not, the first thing is to get them to that position of strength.

When running a thrive cycle:

- Make sure someone really understands their strengths and contributions.
- Give them support to take on more and grow.
- Help them get the recognition they deserve throughout the cycle—not just at the end of it.

It's helpful to keep in mind what cycle someone needs to be in and try to get them to embrace that mindset. High performers who switch roles can try and go directly to a "thrive" cycle without giving themselves time to lay the groundwork. Sometimes people who could be focusing on thriving distract themselves trying to close gaps, and that makes for frustration on all sides.

Growth without promotions

For a while, I worked in a company with no levels and no promotion process. There were—are—some obvious and clear downsides to that, and they came up often.

But here's what I liked about it versus organizations *with* levels: it allowed people to focus on their growth and removed the element of comparison that arises when there are levels. As a manager, it allowed me to encourage people to grow in the way they wanted and was of interest to them, without the temptation of checkboxes. It gave me latitude to reward what the team needed rather than what the ladder said was impactful, which made it easier to push a team-first mindset rather than an individualistic one, and to celebrate and reward people who cleaned things up and made things work better as much as (or more than) those who built new things.

In bad economic times, promotion processes may be more strict, with fewer or no promotions and more people pushed into performance management or canceled altogether. In good times, promotion processes are often events that connect to attrition because they prompt evaluative feedback that may be dissonant to someone's self-image and dissonant with their options in the open market. If review season deems someone not a senior engineer, but they believe they are and can find those opportunities in the open market, then they have the option to maintain their self-image at the cost to your team of one developer.

It's easy to get annoyed or defensive about this; maybe you're right and they didn't meet the expectations, or maybe you're just the bearer of bad news in a process you don't have the power to change. It doesn't really matter; the challenge of any such process is that people feel an absence of control, and as such, they take that control where they can—in a competitive job market, an individual has more control than in a typical review process.

The second challenge of these processes is trauma. People experience them as consistently bad and unfair, and they bring that trauma to any process you enact, regardless of the fairness and objectivity of *that* process. You are almost never engaging with someone's relationship to that process directly; it is invariably mediated by the baggage that they bring from the experiences they have had before.

Once I moved to an organization with a job ladder, I tried to keep the best parts of not having a job ladder: to help people focus on their own growth and view the job ladder as a mechanism of delayed recognition, rather than a set of goals.

Two things can support this:

- Building a DRI mindset
- Aligning with their values

Building a DRI mindset Help someone increase their perception of control by encouraging them to build the DRI mindset we talked about in Section 1, "DRI-ing Your Career". Encourage them to consider their broader goals and their work in the context of what they want out of their life. Identify together opportunities where they can grow and develop and feel good about that, independent of the review cycle. It's likely you have more power over recognition, growth opportunities, and learning budgets than you do over the review cycle—use them.

Aligning with their values We talked earlier (in "Managing Burnout in Others" on page 98) about how often people view their values as a moral good or bad, but generally, it's more nuanced than that. For instance, let's say someone is an exceptional teammate and is resentful because someone who is more self-focused than team-focused advances and they do not. It's true that the value of team-mindedness may actually be holding them back from advancement. However, instead of arguing with the system (where you are the representative of the system), encourage them to *own* their values and the choices they are making. It's possible they would prefer to be who they are than get promoted, and if they admit that, it'll be easier for them to operate from a more empowered place. In a victim mindset, that person might lean into a narrative of "I do everything for the team and it's not valued at all," but in an empowered mindset, they might say, "I value teamwork and I choose to prioritize it, even if it comes at a cost." Admitting both their values and the cost may make them a bit more discerning about what they prioritize—and help them focus on the most impactful teamwork.

Supporting in this way is often about increasing their perception of control, helping them see and focus on their own potential for growth, and supporting them in setting career goals that transcend the organization. The performance review is the scorecard the organization hands out, based on the values and constraints the organization is operating within. While it may have an impact on someone's perceived value under capitalism, it does not define anyone's value as a human being or what they have the potential to accomplish.

Hiring External Leadership

Hiring external leadership is an opportunity to bring different perspectives and experiences into the organization. If you are at an earlier-stage company, trying to grow, hiring someone who has been there before, navigated those problems, and solved them can be a huge plus. As can adding someone who has the other perspective—who knows what scale and a larger org structure look like.

This is the upside of hiring in. There are also downsides:

- Team resentment and rejection are much more likely for new leaders, especially when there are one or more people on the team who think they could have done that job—or even worse, were doing it but poorly.

- Failures are a bigger problem. If an IC doesn't work out, the impact is more limited. When a leader doesn't work out...it takes longer to know, and the impact is more widespread, particularly if they have hired in too. It takes longer to onboard leaders and know whether things are working out, and longer to unravel.

- New leaders will often push for changes in processes and org structures, and that makes sense; needing those kinds of changes is probably at least part of why you hired them. However, new leaders are biased to try and re-create the org structure they just left—something that was built in a different context. This risks creating resentment (the new leader is not seen to be listening or adapting) and being ineffective. Remember that good strategy is context dependent (as we discussed in Chapter 7).

Essentially, hiring in leadership from outside the company is one of the riskiest things you can do. But it can also have an incredibly high impact—both to the team and, if your job is overwhelming and you want the capacity to take on other things, to you personally. The right balance will depend on the team you have, but it should be a balance. If you only grow people and promote from within, you never get that incredibly valuable external perspective. If you only hire from outside, you send a message that this is not an environment where people can grow.

Let's talk about how to derisk your options when hiring a leadership role:

- Developing a good rationale—figuring out when to hire
- Clarifying responsibility—defining scope
- Hiring effectively—what to assess
- Onboarding successfully—setting your new leaders up for success

DEVELOPING A GOOD RATIONALE

The rationale is the reason why you want to hire someone for a leadership role. Maybe it's easy and obvious: someone left and you need to replace them, or you have a team of 20 ICs and can't manage all of them yourself. Even when it is easy and obvious, it can be worth thinking through some alternatives to refine your reason:

- If you couldn't get this hire approved, what would you do?
- If you found the perfect person but needed to wait six months, how would you navigate that?

Once you have gotten creative with other ideas and understood their downsides, you will have refined your thinking such that you can better explain why you want to hire. You may also have some ideas to help you navigate the time it will take to find someone good.

When the rationale is not obvious, it's typically because you're looking farther ahead to what the team will need 6–12 months from now, so it's time to return to your strategy. What does this enable? What does this get ahead of? These things help you manage up to get approval (if necessary), but they also help you communicate down and build team buy-in, reducing resentment and friction.

Finally, there will be a point when you need to explain your rationale for bringing in a new manager to your team. If the team objects to the rationale, often the subtext is that they are afraid for their own growth opportunities. As we've covered extensively in this section, growing and developing people is a lot of work. Having more people who can do that well can in fact improve people's growth and opportunities. This can be a useful line of thinking to incorporate into your rationale.

As such, when you communicate the rationale behind hiring, it's helpful to outline the reasoning into four aspects: the headline, the strategy, the short- and long-term needs and growth, and finally, the forward-thinking, long-term strategy:

Headline: what's changing

> We're hiring an additional two managers for the team to support our ongoing growth.

Explainer: this is how it fits into the strategy

> Our projected hiring is six to eight people this year, which will support two new product areas; as such, we're going to want one manager for each product area to set us up for success going forward.

Short-term needs and growth

   In the interim, Andi will lead Product Area A, and Bobi will lead Product
   Area B—hiring managers for these areas will allow them both to focus on
   the tech lead role.

Forward-looking, long-term strategy

   One of the focuses for these new managers will be to improve our capacity
   for developing people; in time, this will help us do things like support other
   product areas and hire less experienced developers.

At this point, it may also be worth having clarity on how the rest of the team
will be involved in the decision. Will they meet people before they are hired—for
instance, as part of a group sell call? Will anyone on the team interview them?
People will have questions about the process, and part of the subtext will be how
much they should worry about who their new manager might be. The more you
can explain what you're looking for and how it will work, the more you can allay
their fears. You can also use this opportunity to see if anyone has previously
worked with a leader that they would love to work with again!

CLARIFYING RESPONSIBILITY

In my most recent adventures hiring and onboarding (director-level) leadership,
I've found the strongest predictor of success to be clarity about what the job is.
   Possibly that's a product of working in a slightly weird environment with a
"nontraditional org structure."[9] But I suspect it goes further than that. People
generally have a model of what a director does and is responsible for, but that
doesn't map exactly between environments. In product-lead companies, this
looks different than in engineering-lead companies, and nontech companies
are different again. Whether or not you own a budget differs. Whether hiring
is centralized or owned entirely by teams makes a significant difference. The
structure, type, and responsibilities of partner teams (like product and design)
have a huge impact.
   As such, the most useful thing you can do for your new hire—and yourself—
is to get very clear about what they are responsible for and what they are not.
If they are not responsible for hiring, you want them to give feedback and work
with whoever is responsible for it rather than owning the pipeline. If they are
supposed to own the pipeline, you need to get them to own the pipeline rather

---

9 This is a polite way of saying the org chart is very confusing.

than expecting the recruiter to do it. If product owns the roadmap, they will need to provide input, but if engineering owns the roadmap, they will need to *own* it. It is very unlikely that the responsibilities of someone's previous job map exactly to the new one. Every organization has its own idiosyncrasies. When you define the responsibilities, you will clarify those idiosyncrasies and explain the boundaries of them.

People who get into the leadership track are typically driven to take responsibility. This is great and useful. Maybe you are hiring someone to take over everything and figure out the plan; then, you want to be clear on constraints and what outcomes you are looking for. However, if you want them first to own and deliver on their *core* responsibility, and use that to build their credibility such that they can expand outward, you need to make that clear.

I think this is particularly important when hiring Staff+ engineers and clarifying their responsibilities. Staff+ engineers expect to—and are expected to—drive large-scale technological change. In interviews of Staff+ engineers, this is often what they focus on: identifying and understanding what level of opportunity they will have to do that. Maybe in some organizations that is exactly what is needed, but I suspect that almost everywhere it is first necessary to build the credibility and context required to drive that scale of changes forward effectively. As such, making it clear that the large-scale change comes *after* they have onboarded and delivered the high-complexity project is crucial to align expectations and set Staff+ engineers up for success.

HIRING EFFECTIVELY FOR LEADERSHIP

Much of the thinking in Chapter 8 applies to hiring for leadership roles, with a couple of caveats. The first is that it's less necessary to scale leadership hiring than it is to scale IC hiring, which may change your needs and priorities when it comes to process. The second thing is the importance of assessing depth and adaptability when hiring people into leadership roles, so here we'll talk about why that is and how to assess those things.

Depth is important because it helps avoid the trap of "this worked at my last job; now do it again." When someone has depth, they understand not just what was done but why that is what was chosen, what the trade-offs were, what made it successful, and what aspects didn't work that well. They are able to identify value not just in adherence to process but also in the *outcome of the process itself.* When people have depth, they have more flexibility: they have more different approaches and are able to adapt the order in which they institute those changes.[10]

Here are some useful interview questions for assessing depth:

- What is the top thing you want to improve on as a manager?
- How do you think about false negatives in hiring?
- If you could go back in time and undo one technical decision, what would that be?
- Give an example of a long-term strategy that you have deployed to improve something.

Adaptability is about how well someone can read the current environment and adapt their approach to change it. For instance, someone switching from a co-located environment where meetings are very important would need to adapt when switching to a remote environment where written communication is more important. When you hire a leader who is not adaptable, they start trying to force the team to adapt to them. They institute more meetings in an async-driven culture and heavy-handed process in a more chaotic one. The irony is that doing this often makes the team less inclined to shift. When leaders are not adaptable and make self-serving changes, people see through it. When they have pushed through low-value changes because they refuse to adapt, it becomes harder to get agreement on the higher-value changes. Someone more adaptable works with the async culture, picks the most important meeting to add, and gets buy-in for it; goes with the chaos but institutes the minimal process that will have a clear impact. When the team feels that the culture is understood, they are more likely to be open to adjustment.

Here are some useful interview questions for assessing adaptability:

- Tell us about some difficult feedback you received and how you worked through it.
- Give an example of implicit feedback you noticed and how you acted on it.

---

10 Writing this reminds me of the period during which I taught programming in French, my second language. I had enough fluency to explain concepts, but if the first explanation didn't land, I would struggle to find a second way to explain it. In English—my first language—I was able to try explaining things in different ways until I found something that landed with that person. Although I could technically communicate the concept in French, having more depth in English gave me a broader range of communication that better enabled me to meet people where they were at.

- Give an example of a process you changed within an organization. How did you get buy-in? What was the impact?

- Tell us about a change you wanted to make but couldn't. What did you learn?

The goal of assessing depth and adaptability is to ascertain that the person has sufficient understanding that they can apply it successfully in your environment. Take the time to think about the needs of your environment and what is likely to be hardest for an incoming leader, and work those into your assessment process.

SUCCESSFUL ONBOARDING

Once you've found your person, you need to onboard them. Hopefully, the ideas given here help with the core of that:

- Use the definition of their responsibility to make sure you both have clarity about what their job is—and isn't. Help them adapt their skills to the new environment.

- Help them find the balance of adaptability. If they are not being adaptable to aspects of the organization that will not change, you will need to help them realize that and work with things as is. If they are adapting to things that they need to change, push them to assert themselves.

In all onboarding, the time to first win is a critical period. This is the moment when the person demonstrates their value to the team and the organization. In leadership roles, this can be more nebulous, but it is no less necessary. It can be helpful with overcoming residual resistance to hiring in. What the win looks like depends on why you hired the person, but it might look like this:

- Great handling of something complex that doesn't have a clear process (e.g., an incident on a team that doesn't normally have them)

- Overhaul of some critical process (e.g., maintenance or hiring)

- Improved delivery resulting from better organization or clarity

- Improved dev metrics

In one role, I think my first win was leading an incident—ironically, not something I think I'm very good at. But I didn't panic, and I was willing to make decisions and communicate them and provide air cover for the developers who were frantically fixing the problem. I suspect that having someone do those things in a smaller, more intense situation helped people see the value of having someone do those things day in and day out.

Making Change

One of my favorite projects in the Galápagos Islands is how they fixed the turtle population. Humans, helped by the rats they brought with them, had decimated the turtle population, and disproportionately: the female turtles were more likely to be hunted and killed because they were smaller.

A breeding program (*https://oreil.ly/Ho5c7*) was started in 1970. Turtles were bred in captivity until the age of five because if they survive to five, they are likely to survive to maturity. To correct the gender skew, eggs were kept at different temperatures because warmer temperatures produce more female turtles. In 2020, the breeding program for one species of turtle ended (*https://oreil.ly/q_fwL*), going from imminent extinction to a healthy population over the course of a 55-year effort.

Humans generally move faster than turtles, and it's unlikely that you will spend 55 years transforming a team—even 55 months is an extraordinarily long time in tech; 55 weeks is more realistic. But I find it a helpful reminder that change is the result of systematization and consistency, and it's a reminder to play the long game.

It also brings us to consider what we can control—and what we can't:

- We can build great teams, but if we are in a company without product-market fit, people will move on—that's not a bad thing.
- We can't give everyone what they want, nor should we.

There's a positive in the attrition of...

- ...the person who moves on to a bigger opportunity than you can give them
- ...the person who changes their job to support the life they want
- ...the person who makes a lateral move in order to learn
- ...the person who makes a long-desired career change

In our roles, we may have a fiduciary responsibility to an organization, but we have a *personal* responsibility to the people we manage. It's often the case that what's best for the individual is best for the organization, but when it's not, it usually makes sense to part ways. A culture of growth makes organizations more resilient, and individuals...well, we get one life, and most of us are obliged to sacrifice it somewhat to the demands of capitalism, but not entirely.

Software engineers have more agency than almost any other profession; that's why we talk about management so much, because bad management is so much more costly when it's so easy for people to leave.

Galápagos is a special place not because it looks beautiful but because circumstances contrived to make it the place where humans, starting with Darwin, could best understand evolution: the way creatures adapt and change over time. It has not been straightforward, and there are many other outcomes that could have happened. We are lucky to have this reality. As leaders, we have a front-row seat every day to how humans adapt and change and grow. What a responsibility. But what a privilege, too.

Your Action Plan to Scale Your Team

We've covered a lot in this section, so here's a suggested path for making a plan to scale your team. Scaling is not a one-time checklist but a long-term focus that takes time, so make sure you give yourself room to work through it. This list is long, and maybe a little overwhelming, but you're not supposed to go through it all at once—review your next action and give yourself some time to take it, and then come back to it again later.

Step 1: Identify your needs.
What's the strategy? How do you expect your team to evolve and grow?

Step 2: Get your current team in order.
Assess the current state of things and act on the problems you have:

1. Fix nonboarding—if you have any more recent new hires who haven't been properly onboarded, now is the time to start correcting that.

2. Identify any gaps in support and start addressing them.

3. Identify anyone not meeting expectations and start giving feedback and documenting.

Step 3: Address hiring and onboarding.
Add some new people and make sure they can be successful:

1. If your hiring process is an issue, execute "Your Plan to Fix Your Hiring Process."

2. Identify gaps in your onboarding process and address them.

    a. Implement an onboarding buddy program.

    b. Start setting expectations with the team around collective responsibility.

Step 4: Build your bench.
Build out capacity on your team for people to take on more, in a mix of growing people and hiring in:

1. Develop your high potentials.

    a. Identify them using the questions in "Identifying Potential."

    b. Match people to suitable stretch assignments.

2. Identify needs to hire in.

    a. Develop a good rationale.

    b. Clarify responsibility.

    c. Set up an effective hiring process.

    d. Onboard for success!

Section 3 Summary

Congratulations! You are not just a manager but a manager who scales! And also a custodian of an ecosystem!

But...that was always the case. Hopefully, you have more clarity into how that ecosystem functions and what it needs to scale effectively: what do you need to focus on, and what can stay as is (for now)?

In Chapter 8, "Hiring That Scales", we talked about how you can make hiring more effective and more equitable, improving the diversity of your team in the process.

In Chapter 9, "Making People Successful", we covered how to develop people and introduced a model of an effective 1:1. We also covered what to do when people aren't doing well, how to support them effectively where possible, and how to approach performance management, including letting people go.

In Chapter 10, "Building a Bench", we talked about how to identify the people with potential and find opportunities for them, including those people who don't report to you. We covered when it makes sense to hire external people into leadership roles and things to consider as you do that.

Now that you understand how your team ecosystem functions and the levers you have for making it healthier, it's time to move on to what your team delivers. We'll talk about that in Section 4, "Self-Improving Teams".

Self-Improving Teams

There was a period in my career when I was a fixer. I took teams that were struggling, and I helped them improve. Deliver more, deliver better, feel happier.

But what does that have to do with your team? I'm sure most of your teams are probably fine...right?

...How do you know, though?

If I ask you how good your team is, do you have an answer?

...Are you sure?

...What is your frame of reference?

In general, almost everyone thinks their team could be better, *but* almost everyone also thinks their team is mostly OK.

What if I ask you a different question...like...how much has your team improved in the last year?

...How confident are you?

You probably have much more confidence in knowing if your team has improved (or not) than in knowing how good the team is. There are few objective measures of a "high-performing" team. Whether 30 story points (for example) is "good" depends entirely on context. Whether features ship "on time" also depends on many factors, including how you estimate them. Defect rates and incidents depend on definitions and baseline metrics.

A lot of things become clearer when you realize it's easier to measure progress than state. When we ask the question: *how can we be less wrong?*

We can then reframe the question from "How do I know if my team is performing well?" to "How do teams improve?"

If my time as a fixer taught me one thing, it's that there is no one fixer. Turning a team around takes a team. This is an adventure we take together. My success at "fixing" was more success at "facilitation" and team building.

Often in tech we talk about managing software engineers as being like "herding cats." I don't like that metaphor. Cats have no concept of teamwork and operate from a place of pure self-interest. I prefer to think of the work of software leadership as more akin to motivating highly trained raccoons—smart, determined, creative, and resourceful.[1] When they are at their best, they are achieving incredible things.[2] When they are failing, they are often getting stuck in strange places where they had no need to go.[3]

If you want to herd a cat, you are trying to control a creature that doesn't want to be controlled, maybe by tricking it or offering it food. If you want to motivate highly trained raccoons, you need to set the direction, support collaboration, and prevent them from getting stuck. When you herd a cat, you position yourself as the smarter and more powerful being. When you motivate highly trained raccoons, you're setting the direction, but to a certain extent, you're going along for the ride—possibly scarier, but usually faster, and *definitely* more exciting.

It's amazing to me how often, when you ask a software engineer about their best manager, they mention a manager who "stayed out of the way." But perhaps it's to be expected—when managers call people "resources" and operate through control, the ideal becomes the person who does not do that.

In trying to find the middle ground as leaders, we need to understand how to set up parameters such that we don't need to be involved in everything. This is not a section about herding cats. This is a section about setting up direction and ways of working such that the team can get creative and own their own effectiveness—ultimately becoming self-improving.

In this section, we are going to break down four layers of team functioning (and communicating), and we are going to ask four questions in each that will help us identify where to improve:

- Mission (aka the vision)—the why
- Strategy—the how
- Tactics—the what
- Execution—the do

---

1 Also cute and fluffy, unlike most software engineers.

2 Like this legendary raccoon that was photographed riding an alligator (*https://oreil.ly/MXxl2*).

3 Like this one that got stuck in a tank (*https://oreil.ly/5Jk2k*).

This list may seem like it includes categories of action—it does. It's not just the doing of these things but also the communicating of them that ties teams together. Communicating the items on this list plays a major role in scaling teams and leaders. With these things in place and communicated, it's much easier to add people to a team and then entire teams to an organization.

This is not a section about process. We will talk about process, but always remember that process is in the outcomes, not in the action. The secret of process in engineering teams is that we talk like engineers hate "process," but that's not true. Engineers love process. But when they love it, they call it "culture."

This section is about the culture of a high-performing team. It's about how to instill in teams a mindset of growth. Once we understand the layers of communication in teams and how they map to team functioning, we can figure out how to debug and improve teams as well as set up systems and incentives such that they improve themselves.

11

Mission and Strategy

The mission and strategy are how you define what you're going toward and why. They help you directionally orient your team and build a shared purpose and identity.

This is the kind of conversation people often love to have but often resist resolving at the level of concreteness that is actually *useful*. It's easy to get everyone to agree on something like "let's build a great product" and much harder to get that agreement on who the "great product" is for and who it is not for, which is often a pretty fundamental question to answer.

The strategy layer is where your mission becomes real. Determining a strategy is making decisions about how you are going to deliver on the next phase of the mission and how you will evaluate if those decisions are working out the way you expect them to or if they are not.

In this chapter, we will cover how to identify and clarify the purpose (or mission) of your team and develop a realistic strategy to execute on that. Each section includes some case studies showing how these concepts were applied in a particular situation. To give you a broader perspective, I asked some people I admire to tell stories from their experiences.

It's worth noting that the mission and strategy of a team cannot and should not be developed in a vacuum or by one individual. The team mission needs to fit into the needs of the broader organization and be agreed upon with stakeholders. Similarly, stakeholders, like the product managers and designers you work with, will need to be brought into and aligned with the strategy. At their best and most effective, strategies are co-created. Sometimes, your role may be to define the strategy. But more often, your role will be to elicit the pieces of the strategy and facilitate the required conversations such that you can drive clarity and agreement.

Constructing a Mission

The mission is the "why." It's useful at the level where it informs not just what you do but also what you *don't* do. Often when we talk about the mission, we focus on what we are saying yes to, but I would argue that what we say no to is even more important. What we say no to is what defines what we manage to deliver.

CASE STUDY: WORDPRESS MOBILE

Automattic had a mission to "democratize publishing," but it was too broad for any team to take directly, and each team needed to figure out how that mission applied to them. At the time I joined to run the mobile (later Native Apps) team, they were focused on WordPress apps.

The WordPress apps supported *all* WordPress sites: the managed WordPress.com sites, the WordPress sites with Jetpack installed (very similar to WordPress.com but—crucially—not *entirely* the same), and the Wild West of every WordPress site out there. WordPress was built before the iPhone, and even in 2017, it still had a very desktop model—to the point of lacking a proper REST API with authentication. WordPress also had (has) a huge ecosystem of plug-ins that (theoretically) allowed every kind of customization that anyone could imagine. In short, this was the ultimate "works on my machine" situation.

The desktop mindset was pervasive, and many people—both internal to the company and in the broader WordPress community—couldn't conceive of running a website from a mobile device, even though we theoretically had a product to do just that. We would get feedback that without being able to manage (or work with) plug-ins, the app was worthless. In short, there were people who were very confident our mission should be to replicate all of WordPress's functionality on the phone.

The vibe on the team when I joined was not great. The engineers felt put upon and in an unwinnable situation. They were being asked to keep up with the web on two platforms with far fewer people and without normally functioning APIs (which was especially tough for the experienced native app developers on staff). Although the work technically fell under WordPress.org (or at least the GitHub repo lived there), the team was not really part of the WordPress community.

However, we had data and support tickets that showed that we had many people who built their sites on their phones or tablets and used the app to update their sites, even to write lengthy articles. In my support rotation, I saw people who had clearly never even looked at their site from a desktop web browser, resulting in sites that included some errant widget that didn't show up on mobile in the sidebar.[1]

As a result, we instead framed our mission around the second, quieter group of people: the people who showed us what our mission *could* be rather than told us what we *should* be doing. We knew that the hardcore WordPress users, with their default set of plug-ins, their need for all the things all the time, were never going to be happy. But we saw how we were fulfilling a real need for mobile-only or mobile-primary users and that by committing to it, we could make something *great* for them.

So that became the mission: to provide a *great* experience for mobile-only or mobile-primary users.

So when we saw that support data for mobile apps that showed image uploads were causing problems, that was clearly strategically valuable. When we heard the usual complaints about this plug-in or that plug-in not being supported, we ignored them.

I wanted to see how things were improving, so I adjusted my work habits to become more of a "mobile primary" user myself. I got a nice Android tablet and set it up so that I could work on it. I also made more effort to write on my iPad for my personal blog. I started my own mobile-only site, uploading a photo every day to photo.cate.blog. For validation, I mirrored it on Instagram. At first, the Instagram experience was superior in every way. But over time, the app genuinely improved. Even after I left Automattic, I kept posting to both.

To improve people's understanding of the strategy, we would conduct a quarterly "empathy challenge" and encourage people to pick something from a list of options that would connect them more to the people we were aiming to better serve. Options included running some user tests, spending a couple of hours in the support queue looking for trends, writing a blog post from a mobile device, or posting five short image posts from their phone.

---

1 At the time, every new hire started with three weeks in customer support, and everyone did a week in support each year; customer support was a cultural cornerstone of the organization.

During the first empathy challenge we did, an engineer on the team picked the short photo posts. He was so incensed by the experience that he spent the weekend writing a new media picker to fix what he now realized was a horrendous user experience.

Do I want people to make a habit of working weekends? No. Do I want to encourage it? Also no. But I remember that as the moment when the attitude on the team started to shift. The user complaints became more tangible and valid, the mission became more real and motivating, and the victim mentality started to shift.

CASE STUDY: DEVELOPING A MISSION FOR A DISTRIBUTED TEAM

Contributed by Nandana Dutt, engineering director, Google

I joined the Android team in London to work on user privacy. The core Android team was mostly in California. My mission was to hire, grow, and scale a team in London that could quickly develop deep expertise in Android Framework and lead innovations in privacy. High level, the team in London was supposed to be part of scaling the team in California, but it was not clear at the outset what our roadmap would be.

The core team in California already possessed a deep expertise and ownership of various parts of the platform, and the knowledge often resided in individuals. As the most popular operating system in the world, Android has to be nimble to react to trends in the industry or user demands. While the concentrated nature of the knowledge and ownership meant that the core team was well-prepared to perform their standard function, it also meant they were not staffed to support quick pivots or experiments—a crucial skill set for a team that needed to be nimble and responsive to the market. You might think adding more staff members might help, but generally, it did not. Onboarding takes a very long time, and finding suitable candidates is a challenge. To minimize the onboarding time, sometimes people with prior Android experience were hired, but this meant the candidate pool was much smaller.

I realized that a new team taking up strategic priorities from a different location needed to quickly develop deep expertise, and also demonstrate ownership and empowerment to make changes in the core parts of the system and prioritize a set of behaviors that enabled learning, ownership and scaling:

- Having the ability to bootstrap in new areas and develop expertise quickly
- Developing unique expertise in core operating system primitives like process isolation
- Making decisions and judgments that prioritize the user and the ecosystem
- Acting with true ownership
- Prioritizing collaboration and knowledge sharing

As we rapidly ramped up knowledge and established trust and credibility with the core team in California by delivering results, I found opportunities to increase autonomy and ownership. The team developed core expertise in some areas like filesystems, process isolation, and sandboxing. Combined with the strong track record for delivering, the team's roadmap evolved into several key pieces of Android's privacy strategy. The core team's expertise and years of experience were complemented by the London team's nimbleness creating a strong organization. The London team developed a strong roadmap focusing on its strengths and worked closely with the California team over dependencies. Exec sponsorship and explicit organizational support to make the scaling work helped remove organizational barriers.

I grew the team from four members in 2018 to 45 members by 2023. Since the onboarding processes were strong, I could hire anyone as long as they were capable and were interested in operating system work, leading to a very diverse team. We built a healthy team culture based on collaboration, learning, and upskilling one another. I was intentional about building bridges from London to the core team, thus setting up my leads for success. The team was nimble and flexible about taking up new priorities and adjusting to the organization's directions. Today, the team owns some core parts of Android Framework and drives innovative privacy projects in multiple areas in collaboration with multiple parts of the organization.

CASE STUDY: ALIGNING DEVELOPER EXPERIENCE

My final tour of duty at Automattic was running the DevEx team. This team was not in a great state when I joined it; they had some grand ideas about where they could benefit the entire organization, but in reality, the team was a burnout factory because they were in charge of the hiring process for the entire engineering organization, which wasn't going well.[2]

The team was formed by some of the strongest engineers in the organization, the few who were trusted to make hiring decisions. They had ideas about how to make the organization more effective, but this work was just seen as a distraction because the team had impossible hiring goals set for them—work they found tedious and largely unfulfilling. The team was understaffed and unappealing to most developers in the organization, who preferred to ship products rather than review code tests. Being under so much pressure was corrosive to empathy, and as a result, the team often brought the worst part of hiring (evaluative judgment of others) to the other things they were interested in—delivering drive-by comments or pushback at the end as they didn't have time to get involved earlier and educate or build consensus. It also pitted the team against other teams. Any team that wasn't delivering would blame the DevEx team for not hiring enough engineers.

To shift the team, we had to extract ourselves from the hiring hole (more on that in "Case Study: Scaling Hiring" on page 280), but beyond that, we needed to embrace a mission that aligned us with the teams we worked with rather than putting us in opposition. We also needed a mission that could be proactive (rather than reactive) and supportive (rather than judgmental).

We ended up with: be a support for the entire engineering org, focused on acting as an accelerant at the inflection points in teams and individuals.

There are some keywords here. The first is *inflection point*. Do we invite ourselves to weekly meetings? Absolutely not. We defined *inflection point* around significant team changes, such as new team members, new leads, newly formed teams, or changed focuses.

The second keyword is *support*. Do we dictate actions? Vehemently no. *Support* means empowering the team members, coaching, providing resources or active help where possible. One way we were able to create value was to coordinate training and support, like organizing groups of people to go to LeadDev events. Team leads were often putting themselves

---

2 The hiring philosophy of the company was around using paid "trial" periods working on the actual product, which had been inconsistent, often unreasonably long, and generally a poor candidate experience. The team had moved to something they called a "synthetic trial" as a result, which was more like a long but somewhat collaborative take-home code test—a meaningful improvement but insufficient.

last, overwhelmed, and finding it impossible to make time. By giving them a concrete opportunity to prioritize their own growth, normalizing that and making it just a little easier, we were able to provide higher impacts relative to time spent.

Alignment around a broader mission helped us expand the team. Over time, platform teams were added with their own specific points of acceleration (like releases or through the benefits of componentization). And it allowed hiring to remain important but be part of a more holistic—and rewarding—whole.

While there were some intractable problems in this situation that a better mission could not (and did not) solve, having it was part of having something to go toward rather than a situation to escape from.

DETERMINING THE MISSION

The mission needs to be aspirational but also within grasp. In the WordPress Mobile case study, we took an existing group of users that we could serve *better* rather than an entirely new segment. In the Android example, Nandana shared how she built out a meaningful mandate for a new team. Similarly, in the DevEx example, we clarified the aspiration of the team to make it more tractable, rather than coming up with something entirely new.

Often the *where* is handed down, but it's on you to extract and articulate the *why*. You have to craft a sentence that captures the purpose and guiding principle of your team–something that captures not just the proximate objective but also the guiding idea that determined it as well as some sense of what will come after that.

The value of a mission often gets overstated, largely, I think, because founders love to talk about "the mission" and "buy-in to the mission." That kind of rhetoric speaks more to the value of a company mission as a sales and recruiting tool. At a team level, the main benefit of a clear mission is in better decision making. Teams without a mission often spiral and spin from thing to thing... but having a mission doesn't mean much if your strategy, tactics, or execution is lacking.

Determining a Strategy

Strategy is the "how" and is made up of proximate objectives, or next achievable goals.

I suspect strategy sometimes gets a bad rep, because "strategic" leaders can often be all talk, no impact. Meanwhile, people who are less talk, more impact often get deemed "not strategic" or "execution focused"—with great execution, their strategy seems obvious, leading to it being undernoticed and undervalued.

In Chapter 6, we talked about the role of strategy in management. Here, we'll talk about strategy in terms of team execution and product development. Again, we'll turn to Richard Rumelt's fantastic book *Good Strategy/Bad Strategy*, truly one of the most useful books I have ever read. It fundamentally shifted how I think about strategy and gave me a workable definition, one that balances the headlines and the communication with the actual value that gets delivered.

> *A good strategy does more than urge us forward toward a goal or a vision. A good strategy honestly acknowledges the challenges being faced and provides an approach to overcoming them. And the greater the challenge, the more a good strategy focuses and coordinates efforts to achieve a powerful competitive punch or problem solving effect.*
>
> *Bad strategy tends to skip over pesky details such as problems. It ignores the power of choice and focus, trying instead to accommodate a multitude of conflicting demands and interests. Like a quarterback whose only advice to teammates is "Let's win," bad strategy covers up its failure to guide by embracing the language of broad goals, ambition, vision and values.*
>
> **—RICHARD RUMELT, *GOOD STRATEGY/BAD STRATEGY***

As we talked about in Chapter 7, good strategy requires depth and breadth. It takes into account the context of your team, organization, and the broader world.

Unfortunately, most strategy is bad.

CASE STUDY: SCALING HIRING

When I took over the DevEx team at Automattic, the goal was to hire one hundred developers a year.

This was three to four times the number of developers that had been hired the previous year. And the number of people working on hiring had been cut in half.

The book *A Beautiful Constraint* by Adam Morgan and Mark Barden (Wiley) presents the idea that when you change a goal fundamentally, it forces you to change your approach fundamentally—that is, if you aim to increase growth by 10%, your ideas are framed about what you're doing already but *more*. If you want to achieve 10x growth, then you have to take a fundamentally different—and more creative—approach.

So, with a goal of hiring four times the number of people to the team than ever before, and only half the people working on hiring, and also then needing to onboard new people...the problem had dramatically shifted, and continuing with the same approach was clearly not viable. In addition to needing to hit an increased output of three to four times as much as previously, we needed to onboard new people into the hiring work itself—and address the reasons why there had been so much churn there. We were in the 10x territory.

Step one was to make the goal feel more tractable. Framing it as hiring one hundred developers felt so far away as to be impossible, which discouraged the whole team. Further, talking about a large number goal to a team that had never met a number goal was meaningless. Since the team never met the overall hiring goals, nothing was really changing other than by how much the goal was missed.

Hiring inherently is an uncertain process. People drop out, accept competing offers, or take the process more slowly because they have things going on. A raw number of hires is also not necessarily the right metric because if you achieve the goal but those people are not successful, you have achieved a number at the expense of the organization (and at a cost to the careers and lives of a number of people).

We reframed the goal to make it a much more achievable goal of hiring two people per week. This changed the problem to one of throughput. We modeled our system, monitored pass rates, and then set additional subgoals:

- Number of interviews a week
- Number of code tests a week
- ...and so forth

This allowed us to identify key pieces of the process and work on scaling them. If we needed to do 20 interviews a week, we needed to build that capacity and train around 15 interviewers who would do one or two interviews a week (allowing for time off). To train that many interviewers, we needed to identify and train people who were capable of training other people. We followed a similar strategy with code tests.

This modeling also allowed us to identify the interdependencies in the process. A lower pass-through rate at code test was traced back through a slight reduction in pass rate at interview to become a problem of resume-review calibration. The resume calibration was also causing other issues: overfilling available interview slots, which made people book further out, resulting in a high no-show rate.

Having broken down the goal in this way, it was easier to track progress. Long before we consistently hit two people a week, we could see the progress we were making on the constituent parts of what it would take to get there. This helped the team maintain optimism even when things were difficult. And later, even after I had left, when the team needed to scale up again, the model was clear. By 2021, the team was consistently sending out three to four offers a week and was on track to hit a goal of two hundred per year.

The old process had a lot of one-to-one dependencies that we wanted to remove as much as possible. For example, sending emails that introduced one person as an interviewer was replaced by a Calendly with the availability of all the interviewers. We added other automation (via a chatbot known as "friendly raccoon"[3]), developed clear rubrics, and set up systems to close feedback loops (so that an interviewer who advanced a candidate could learn how the rest of the process went).

Running a scaled process required investments in process—particularly to support the core part of the strategy, which was to move from a few people doing a lot to many people doing a little—and manage associated increases in overhead:

- By making many people do a little, inherently the work became more manageable, and we increased engagement and enjoyment. By removing one-to-one dependencies, we were able to dramatically reduce bottlenecks and so reduce time in process. Anything incoming got allocated to someone with availability, whether it was an interview scheduled from an interviewer pool or a code test allocated to a pool of code test reviewers.

---

3 Developed by Nikolay Bachiyski, although my totally normal level of enthusiasm for raccoons *may* have influenced the name.

- We removed the ambiguity. Instead of intense 1:1 training, we adopted consistent rubrics throughout and calibrated people against them. Every interview covered the same points; every code test evaluated the same things. This changed the question "What do you think?" to "How does this interview/submission evaluate against the rubric?"—a much easier question for the interviewer to answer. Even better, this reduced subjectivity, making the process more fair and consistent and improving confidence in the evaluation.

- We shifted the culture to make it clear that this work was visible and valued. We provided ongoing training and calibration, creating "review" groups made up of people who were particularly great at interviewing or at reviewing code tests. But beyond providing support, we worked really hard to make sure people knew this work was *appreciated*. We adjusted and added process to support a culture of appreciation: crediting people involved in any new hire and sending out quarterly tokens of appreciation, from custom swag to handwritten thank-you cards.

This example is one huge process with many subprocesses, but the individual details of many of them are not that interesting. What was interesting was the North Star of a scaled and sustainable process and the core idea of many people (ideally, everyone) doing a little. Tying all subprocess back to this and ensuring that it supported it was key. When I consider the missteps we made, the majority of them stemmed from forgetting this North Star and being reactive or succumbing to pressure.

CASE STUDY: ONFIDO STUDIO

Contributed by Yuelin Li, Chief Product Officer, Onfido

At Onfido, we provide digital identity verification that is used by companies globally where verifying a person's identity is important or a regulatory requirement, like in financial services.

Over time, this has become more complex, as more potential methods to do verification develop, and finding the balance of convenience and security for each user while complying with geography-specific regulations continues as a concern. While initially, these methods would be pieced together by companies trying to verify the person's identity, over time, there has been a shift to provide not just specific aspects of identity verification but also identity orchestration platforms that allow greater flexibility and optimization.

Given these shifts, we reached a point where we needed to decide whether it made sense to shift our approach—either focusing on distributing to aggregators who offered an identity orchestration or building an identity orchestration approach ourselves. We opted to shift to creating our own identity orchestration platform (named Onfido Studio), and we developed a strategy to enable doing so.

Some key aspects of this strategy:

- Focus on engineering and product design capabilities to produce a great user experience that allows users to efficiently verify their identity and move on to whatever it was they were aiming to accomplish—like opening a bank account or renting a car.

- While we staffed a platform team, we also worked to shift the mindset internally to a more platform-based approach. This "studio first" mindset encouraged feature teams to consider their work as part of the orchestration platform beyond just the feature itself.

- Practically, this meant doing things like creating documentation so that product development teams could build their own integrations as part of their development roadmap.

  At a higher level, we aligned incentives by working cross-functionally with customer and sales organizations to set joint adoption and migration goals.

- We also created an expanded product offering, based on the orchestration capabilities, such as localized compliance packages for each core market.

A year later, we launched our Onfido Studio product as part of our Real Identity Platform offering. Adoption of the platform accelerated as it became more feature-complete and differentiated, and the company is on track to move (almost) all clients onto Studio within three to four years from launch.

These are my greatest learnings from the experience:

- In the B2B SAAS context, getting large customers to migrate onto a new platform will take time and need to be carefully managed, so it's important that it forms a core part of company strategy rather than just product strategy.

- It helps a lot if the new product is very cool and differentiated and really helps them do things they weren't able to do before.

CASE STUDY: BUILDING THE MOBILE INFRASTRUCTURE TEAM AT AUTOMATTIC

Contributed by Lorenzo Mattei, director, Native Apps, DuckDuckGo

When I joined Automattic, I was the first mobile DevOps engineer in the mobile team, and I found a very interesting situation: the product teams owned everything related to the development lifecycle, from feature development to tooling to releases to quality. There was also a common sense that "things" should be improved given the growing team, number of products, and their complexity but no clear plans or priorities.

With a defined mission of improving development velocity, I needed to define some strategies, and I started to collect the main issues, pain points, and inefficiencies perceived by developers and by their managers and to do my own research. The outcome was a pretty long list of topics:

- Long release process
- Lack of automation in the release process
- Poor development tooling
- Slow CI (continuous integration) with low effectiveness (too few tests and features)
- Low app modularization and too-complex code sharing flows
- Inefficient testing and quality flows

With this list of areas of needed improvements, the next step was to define a reasonable roadmap to address each of them.

While every DevOps engineer would like to be able to build their full vision of the developer experience up front and then deploy it as a finished product, this is rarely possible. In fact, it's rarely the best thing to do, especially when you are starting from scratch and have quite a long project in front of you.

A much better option is to incrementally roll it out, test, get feedback, and iterate. This way, developers start to benefit from the incremental improvements much earlier, and their feedback helps validate your direction.

Fine-grained incremental rollouts, though, are generally challenging in this kind of context because every time a change is deployed, it's important to make sure every user (i.e., everyone in the team) is aware of and comfortable with using it.

In our case, we also faced some additional complexity:

- Every product team was growing pretty fast, and in a context where developer tooling and processes change fast, onboarding new people is more difficult and slower.

- We had a goal to build a shared infrastructure across multiple products. We were starting from a very fragmented situation where processes and tooling were different for every product, and, sometimes, even inside the same product line, they were different across platforms. This meant the "diffs" we shipped were not the same for every instance, and we had to account for this while deploying and evaluating feedback.

- In a globally distributed company, there are no downtimes during the day. There's always someone in the middle of their working day. This meant we had to pay particular attention to communication to ensure minimal disruption and maximum visibility to everyone, independently of their time zone.

While everything previously mentioned is not uncommon for DevOps work, it still requires a good amount of coordination, which looked like quite a big overhead for the fast iteration process we had envisioned.

To derisk and make it more sustainable, we opted for a long-term three-step strategy:

1. Initially, the mobile DevOps team takes full ownership of the areas to work on.

2. As a full owner, implement the changes, test them with the contribution and feedback of the whole team, and iterate.

3. When ready, give back ownership to the product teams when it helps the team ship faster and be fully autonomous in their daily work.

Implementing this strategy couldn't be a one-person-band effort. We needed a team that could own critical steps of the software development lifecycle, like releases and quality, while working on improving it, so we started building the Mobile Platform team while leveraging Automattic's incredible experience in distributed work to make a team with people across almost all time zones work effectively.

Over my four years leading the team, we iterated on the tactics, but the main strategies proved to work well, allowing us to make strong improvements across all the areas and to support the growth of the company in terms of people and products.

As a first step, we started taking ownership of the most critical areas. Sometimes, it was an actual handover from the product teams—for example, with releases. Other times, we just started working on an area that was not owned by anyone before.

Then we started to work on improvements, based on the original analysis, developer feedback, and our own direct experience as area owners. Since our strategy helped us remove much of the overhead previously outlined, we aimed to work in small and fast iterations. Our team was the first tester and feedback giver, and only when the change was of a meaningful size and well tested internally did we deploy it to the product teams. We aimed to present and discuss every change that could have a high impact on a team with them before the deployment. We also tracked how it worked for us, which turned out to be useful to showcase and support the next iterations of the evolving process. Overall, this workflow allowed us to proceed with our plans with good velocity and without disrupting product teams' work.

When I left the company, we had reached the third step of the preceding strategy in all areas. For example, we implemented an effective release process and related automation, and we were in the process of giving the responsibility of handling releases back to the product teams. Over time, we learned that reverting ownership was the next step to increasing their velocity and productivity. They had everything they needed to ship on their own, but coordinating with additional teams just slowed them down.

Outside of releases, however, the mobile DevOps team decided to keep the ownership. This allowed us to continue to leverage the wins of a centralized point of view. As a team that worked across all products and business units, we were uniquely positioned to be able to share code, process, and infrastructure, which maximized the opportunity for everyone else to take advantage of those centralized improvements learned from all the other teams. When making changes to infrastructure, it's important to consider what should become centralized, such as process and related tooling, against what should go back to product teams, such as executing releases.

The main goal of a mobile infrastructure team is to help product teams ship faster and with higher confidence. Through this work, we built a streamlined and automated release process. We iterated on the testing processes and moved from a classical quality assurance approach to a more effective quality engineering one. We improved code-sharing practices, tooling, and developer experience. We iterated on our CI to the point of building our own CI infrastructure to ensure fast, reliable, and wide-coverage support to the developers.

As a result, developers could rely on CI for a number of tests and checks (which, at the speed the company was growing at that time, was particularly important for new members to merge confidently), product teams could ship quickly and reliably, and users got timely and high-quality updates for all apps.

Developing the Strategy

It's crucial to note that strategy taken from one place won't work elsewhere; there are inevitable variables that you can't control. The type of strategy you use to run a small team of senior engineers effectively is not the same one that will work with a large team that skews junior. The strategy that effectively determines product-market fit for a consumer product is unlikely to work as well for an enterprise product working on slower sales cycles. There may be key tactics that work effectively in both situations, but adopting everything wholesale is unlikely to work.

This can be true even within your organization. **Everything** is contextual. Before I had to scale hiring across the organization, I had to overhaul the hiring of mobile developers. Many things were reusable, but some key things were not. A specific example: on the mobile team, I was able to train everyone myself, but that fell apart very quickly for a bigger process, and as a consequence, things like rubrics took on greater importance more quickly.

Strategy needs to be tangible such that progress can be measured. When we decided to improve the media experience on the WordPress apps, we made concrete improvements identified based on user feedback, looking at other best-in-class media experiences. Then we measured, with metrics around media upload and support volume. By making scaling hiring more tangible, we could set metrics around building out capacity, such as how many interviews were conducted in a week or how quickly someone could schedule in. Understanding the top-level metrics (like number of people hired) is important, but when you have lagging metrics, leading indicators (like number of interviews or code tests) are critical for closing feedback loops sooner and allowing you to identify problems and get ahead of—or at least mitigate—them.

If your mission is where you are going, your metrics are what tell you where you are. Ask these questions:

- What do our metrics tell us about what is working?
- What do our metrics tell us about what is not working?
- What are our leading metrics?

Now develop hypotheses that will help bridge where you are today and where you want to go:

- Who is your target audience?
- What will help you serve them better?
- How will you know it's working?
- How will you know it isn't working?

Making the Mission and Strategy Work Together

Strategy is the decisions that are made to deliver on the mission. If you haven't made the decisions, you probably aren't working toward the mission effectively. Teams without a strategy struggle to make decisions. The mission is deliberately broad enough that it is open to interpretation. A number of things can support it, allowing people to advocate for and justify their own pet projects. The question of what supports it and how much requires some meaningful thought and data.

The most important way in which mission and strategy work together is by defining what you decide **not** to do. Some things you might decide not yet, and some things you might decide not ever, but nondecisions are decisions to fail. You need to look at the different ways to make progress and make decisions, knowing what kind of results you want to see and when to validate or disprove your strategy—and when a failing strategy ultimately disproves the mission.

In software, you can typically model how your one-week retention influences your one-month retention and your annual renewal; there's a great case study from Duolingo (*https://oreil.ly/WunpO*) on this topic and how they finally broke through a period of stagnating growth. This doesn't mean that I'm advocating for tracking every element of user behavior; having worked in the privacy space, I know that is not necessary (and question whether it is even helpful). It is about building a holistic understanding of the key metrics that matter and ensuring that they fit in with the overall picture of what the team—and business—is trying to achieve.

In product development, I have found it helpful to theme things: grouping smaller projects together to give them a sense of coherence and some metrics that indicate that the theme of work has resulted in improvement. This allows teams to feel a better sense of ownership and impact.

Finally, the strategy has to fit the team you have. You might have 10 initiatives, but if you can only staff two...then you have to choose two. Doing everything is not a strategy.

Strategy is not static. It is the product of an assessment of the current circumstances and best path forward. Mission is more constant, but the mission should also change over time: the mission that helps navigate finding product-market fit is distinct from the mission that helps you scale and deliver. It's worth thinking about and understanding what would make you change your approach. Organizations that shift strategy too frequently fail in chaos, but organizations that don't shift strategy enough fail via stagnation. Commit, validate, and either shift or recommit.

Themes of Work

This is a technique to build a roadmap that clarifies your strategy and situates projects within it. Depending on org size, you pick the set of things that you can work on and tie that to the impact you want to have and the metrics associated with it.

For each theme of work, you need to explain the following:

- Why this theme of work is important
- What are the proximate objectives—this is how we know this theme of work is working
- What's in flight
- What's next
- What's finished

Within the theme, you can list the specific projects that are being worked on. For each one, include:

- Why it is important
- Expected impact on proximate objective
- ETA

If you repeat this regularly, you can revisit your ETAs with actual delivery and use that to progressively be less wrong about estimation.

An example:

Theme
Increase recurring revenue (subscriptions) by X% (Table 11-1)

Why
Required for a sustainable business and the strongest indicator of product-market fit

Proximate objectives
Increase new subscribers by Y%; decrease churn by Z%

Table 11-1. Example projects in the recurring revenue theme

Project	Why it's important	Expected impact	ETA
Add paywalled features and upsell prompt to settings screen.	User research demonstrated that people don't understand what is in our paid plans; this is an opportunity to show them as part of their workflow.	Settings screen is opened XX times a day, with a YY conversion rate that would result in ZZ new subscriptions a day.	W/C MM DD[a] *Note: smaller project, estimated to the week.*
Implement subscription management platform.	Subscriptions were an add-on to a platform built for one-time charges; as such, they are not working well, and our renewals are well below industry standard because of poor handling of things like expired credit cards.	Bring renewal rate (currently XX%) in-line with industry average YY%, +ZZ%/year.	MM *Note: larger project, estimated to the month.*
[a] W/C = week commencing			

Theme

    Improve application stability (Table 11-2)

Why

    App is unstable, causing data loss and negative app store reviews

Proximate objectives

    Bring Android app health in-line with comparable apps, according to Google Play Store data

Table 11-2. *Example projects in the application stability theme*

Project	Why it's important	Expected impact	ETA
Investigate and address top five crashes.	Crash rate is too high.	Top five crashes make up 75% of total crashes; addressing these will bring us in-line with comparable apps according to Play Store data.	W/C MM DD.
Refactor CoreApp View	Most data-loss issues have resulted from the CoreApp screen, which is >5,000 LOC[a] in one file and, as a result, hard to reason about or fix. By migrating this to standard MVVM[b] architecture, we will be able to add unit and integration tests and fix many previously intractable issues.	Currently, XX% of support tickets concern the CoreApp View; reduce by YY% ~immediately following the refactoring and by ZZ% by EOY.[c]	W/C MM DD

[a] Lines of code
[b] Model-view-viewmodel
[c] End of year

12

Tactics and Execution

Tactics and execution are the movement toward the goal. They help you surface the state of work and keep people aligned and productive. Tactics and execution are the foundation layers of team functioning.

Tactics is your process layer. It's rare to find neutral emotions about process, and it's often a source of significant arguments, sometimes understandably so given how it affects people's day-to-day. Some people (I call them "Process Monsters") love process *too much*, and others don't take the impact of process seriously enough. Personally, I believe the judicious application of process is a superpower, and Process Monsters can be incredibly useful when they can channel themselves (or be channeled) effectively in a given context. They can also be incredibly destructive if they focus on the wrong things and don't prioritize outcomes.

At the execution layer, the ideas and theories become *reality* and get in the hands of your users. The impact of the team is what the team ships.

In this chapter, we'll outline the different layers of team functioning and how tactics and execution fit with the mission and strategy layers we discussed in Chapter 11. Then, we will cover developing team tactics (or processes) that help teams have clarity and work together more effectively, and finally, how that ties to execution.

Defining the Layers

The first lesson of good communication is often to "know your audience," to tailor your communication style to the people and situation. Effective team communication happens at multiple levels of fidelity and detail: a focused mission statement, a clear strategy, shared tactical agreements, and, at the highest level of detail, the day-to-day execution. Each of these layers of communication is tied to some level of functioning, from directional alignment to individual productivity. We need all of them, and they all need to work together to be coherent.

THE MISSION

Personally, I hate the word *vision* because it has undertones of delusion,[1] so I use the word *mission* instead. An effective team has a mission they can cluster around. It provides a guide for what to take on—and what not to. As we talked about in "Case Study: WordPress Mobile" on page 274, when I ran the mobile team at Automattic, the goal was giving mobile-only and mobile-first WordPress users a great experience. Having this mission gave us something to aim for and a sense of which feedback was important. It also connected the team to the company's broader mission to "democratize publishing."

If the mission is missing, the team risks falling into analysis paralysis or abdicating decisions entirely. It's also just demotivating, making it harder for people to understand the impact their work is supposed to have. In the absence of some guiding principle, it's easy for people to pick and choose the data to support their opinions. Abdicating decisions entirely can involve emphasizing the way the work is done rather than the work itself: either drowning in process or emphasizing a "culture" that is by its nature unsustainable and unrelated to any meaningful goal—for example, developer-driven discussions around tooling that has no clear path to monetization or a focus on individual output without considering what that output rolls up to.

THE STRATEGY

It's not enough to have a mission: we need to have a strategy that pushes us toward it. Strategies are proximate objectives that support the mission. For example, we may want to "deliver a sign-up flow that allows people to create a mobile-optimized website from their mobile devices" or "improve the media experience, resulting in more people uploading more media." These strategic objectives can be owned by subteams.

If the strategy is missing (or lacking), then the case can be made that almost anything supports the "mission," creating overwhelm, indecision, and conflict. The mission is explicitly broad and (hopefully) inspiring. The strategy is where you articulate what you are going to do to drive it forward and what you are not going to do (yet, or ever). Having an explicit strategy allows you to decide how

---

1 You know who is always operating on a vision? Supervillains, and also billionaires as they destroy your favorite social network.

the team you have can best move toward the mission. Writing it down, revisiting it regularly, and using it in your communications about what you're doing—and when you emphasize what you are *not* doing—helps make it real.

THE TACTICS

Tactics (or process) turn strategy into something that individuals can deliver on. They break down how work and communication are managed across teams. Adding this layer without having a strategy risks an elaborate performance of process in which most people will not see the value. This layer must support, not overpower, the strategy.

If both strategy and tactics are missing, it may be tempting to start with tactics, but you can't go too far without the missing strategy layer previously mentioned. On teams I've led, this has typically involved things like how we define features, how we measure performance of new features, how we make architectural decisions, and how we plan out and report on the roadmap within the theme of work that the strategy lays out.

If tactics and process are missing, then there is a huge overhead to any kind of coordination, and status is challenging to find (and maintain). No one knows what is going on, and whether or not they feel like it's "good" rests entirely on emotion—the most inconsistent and hard to measure of "metrics."

In the mobile team example, when I came in, I was told the problem was that the team wasn't delivering and wasn't focused on the most important work. However, when I looked into it, I found that the projects were (mostly) driven by user value and did align with the needs of the business. The actual problems were (1) that wasn't being communicated, (2) too much of the team was working on huge projects that took a long time to deliver and show value, and (3) the work wasn't well scoped and broken down in ways that would demonstrate incremental progress.

Putting in processes to address how we talked about projects and the progress that was being made on them improved delivery as well as the perception of both the speed of delivery and the value of what was being delivered. Changing the spread of projects so that there was a better mix of large, medium, and smaller changes being worked on improved overall momentum. Setting up systems that were more responsive to user feedback with a consistent overhead (i.e., an allocated maintenance rotation rather than whoever—if anyone—is willing to fix something) helped balance that kind of work with the big projects, creating a better impression around the team being willing to address incoming feedback.

THE EXECUTION

The perfect mission-aligned strategy, even perfectly managed, still needs to be executed. This involves the day-to-day work and communication around it. It includes things like standups, the way new task requests are made, and how PRs are reviewed and merged. Having a lot of execution without the other pieces results in a lot of churn and detail-level activity that doesn't roll up into a coherent whole.

When execution is missing, it's obvious because very little happens. Projects may be well defined and communicated, but they don't move. The strategy may be clear, but no progress is made toward it. Everyone may buy into the mission, but it doesn't really matter when so little is getting done. We'll come back to execution later in this chapter.

META

A common phenomenon I've observed (and heard about) in startups is a mission that rallies people and a focus on individual-level execution. The strategy layer might exist inside some hive mind of senior leadership, but it isn't communicated regularly or consistently. The tactics are not well defined or are focused too much on an individual level (a common pitfall of companies that think they can survive without the "overhead" of people management), so teams struggle to communicate and scale.

It's possible to unlock tremendous momentum and improvement by adding the strategy and tactics layers to an organization operating without them. The pitfall is that when organizations don't have them, it can be because someone in a position of power doesn't see the value of them. This can lead to stalling and arguments as newer leaders try to "build alignment" or just wholesale adopt the processes they liked from their previous workplace. These arguments tend to reinforce that mindset, that those things are in fact a distraction from the real value of "getting sh*t done."

I've seen this failure mode up close in different forms. The executive who believed their value was as a strategist and, after more than a year of arguing, got shunted to one side. The design operations person who brought in so much process (theoretically tactics) that it ground the design org and ultimately product delivery to almost a complete halt.

I watched both these situations play out in frustration, annoyed at the impact on my teams and the broader organization. With time and distance, I could see the valid points. The strategist was right; there *was* a lack of strategy. The operations person was correct; there *was* a lack of quality and consistency in design. But going all in on a solution when others hadn't agreed on the problem was ineffectual and—worse—undermined the ideas they were trying to put in place, rendering meaningful improvement harder.

In leadership, there are some things we need to fight for and some things that it's worth it to push through against organizational resistance. But we should be very careful and deliberate about what those things are. If the fight is over everything, the result can easily be nothing. And bluntly, that's unnecessary. Although it can be frustrating (and tedious) to come into an organization that has wildly out-of-date practices and try to drag them into the current age of product development, change management is relatively well understood. It is hard in the way math is hard, not hard in the way that quantum mechanics is hard—with some time and patience, most people can achieve a level of competence with it such that they can be effective.

MAPPING

We covered strategy and mission in Chapter 11, and next, we'll place them together with tactics and execution (which we will cover in the second half of this chapter). You can think of the four levels of communication as follows:

Execution
> How individuals work

Tactics
> How teams work and what enables them to work together

Strategy
> What the organization delivers

Mission
> The guiding principle (or North Star) of the organization

Or:

Execution
> Today/this week

Tactics
> This month/this quarter

Strategy

Next quarter/this year

Mission

Next year/indefinite

These levels build on one another in both directions—the goal is not top-down or bottom-up; it's balance. To be effective, leaders need to understand that improvement needs to happen at all levels concurrently to achieve that balance.

Functional, effective teams have harmony among all these things. Dysfunctional, ineffective teams have confusion at one or more levels. The irony, in my experience, is that often the concepts are there at least somewhat, but the consistent communication—the tying them together—is not. It's that absence of clarity that is allowing disconnects and disillusionment to grow and grow.

Developing Tactics

Tactics are the "what," or the process layer. They are about the way teams work and the APIs that enable them to work together. They define the delivery milestones of the strategic objectives.

This layer is what makes the connection between the strategy and the work of the individual. It's how people understand how their work fits into the bigger picture. It's also how teams become interoperable, being able to work toward a common goal without undue overhead.

It is critical that tactics support strategy. Tactics without strategy is just busy work. Good process is in service of an outcome. Think of the concrete activities that make things go (good process) versus rituals performed with no clear purpose (bad process).

Teams without tactics struggle to work effectively, particularly as organizations scale; collaboration is hard and only gets harder as more people are involved. Tactics are like your team API; they serve to manage your integration points between teams or up.

When determining tactics, think about:

- What do you need to know to be effective?

- What do you need to measure?

- How is this problem generally solved?

CASE STUDY: THE ACCIDENTAL INTRODUCTION OF DEADLINES

When I took over the mobile team at Automattic, I followed my usual strategy for understanding a new team: people, projects, process. In the projects portion of this, I tried to take stock of what projects were going on, what they were supposed to achieve, and what timeline they were operating on.

This took some time but resulted in a document that outlined the current roadmap (in a similar format to "Themes of Work" on page 291), which in the spirit of open communication, I published on the internal team blog. I tied all the projects to their actual strategic impact, so things like "networking refactor" became "full support for Jetpack sites."

As part of this, I asked a question that seemingly had not been asked on the team prior to my arrival on it, which was "When do you think this will be done?"

And this is how I inadvertently implemented deadlines. Of course, we overshot all of them at first, but we did get progressively less wrong over time. Part of that was putting this document together regularly, outlining what we were working on, why it was important, and what we hoped to achieve in the next period.

When I handed over the team to someone else, the early feedback she got from the team was that this was a pointless exercise and she should stop doing it. Her response? "This isn't for you; it's for me." It was a deliberate and recurring event that forced her to take stock of what was going on and where there needed to be intervention. Other leaders in the organization found it useful for their own awareness and planning, as it allowed them to have an overall understanding of what the team was doing without needing to get into the details. When I moved to another team, it was one of the first things I implemented, and adoption continued to spread across the organization.

CASE STUDY: PEOPLE AND PROCESS

Contributed by Jill Wetzler, VP engineering and consultant

I joined Lyft in the summer of 2015, just as the hockey stick of headcount growth was about to surge. My team, then known as DevOps, consisted of seven mid- to senior-level engineers, most of whom had been at the company long enough to be well known and comfortable. DevOps was a bit of a misnomer—this team was responsible for all server provisioning, security and secret management, the full observability stack, CI/CD (continuous integration/continuous delivery), and all services and tooling related to developer environments, among many other things. Not only was this a vast amount of surface area, but all of these services were critical infrastructure pieces for every engineering team at the company in a year when headcount was about to triple. Despite the DevOps team's best efforts to prepare for scale, things could and would break in interesting ways, putting this team on the front lines of support requests from other software engineers on customer teams.

Most teams have those one or two individuals who just really love to help others. If they could, they'd spend their entire day talking to people with interesting problems, answering their questions, getting them unstuck, and then reveling in the warm, fuzzy expressions of gratitude. On my team, we'll call those two individuals Samuel and Wes. In my early 1:1s with team members, I sought to understand the team's current pain points and anxieties. I was looking for quick wins—things I could do as a new manager that would be immediately impactful and help me build trust quickly with the team. The support load was already causing strain, especially with Samuel and Wes. Wes, arguably the most knowledgeable on the team, described the situation as "People run into problems with their dev environment or their service graphs or their tests, so they DM their favorite DevOps member on Slack and ask them for help." As new members joined these customer teams and ran into issues during onboarding, they were also encouraged to send DMs. Their onboarding mentor's favorite DevOps engineer became their favorite DevOps engineer, and the support load grew as more and more people individually requested help. This approach to internal dev support was a powder keg sitting next to a match that the Recruiting team was in the process of lighting. We would not be able to adequately scale a process that relied on individualized support.

The medium- to long-term plan was already self-evident. We needed to add people to the team, we needed a plan to split the team such that we could reduce any one individual's area of responsibility, and we needed to invest in things like onboarding, documentation, and usability. But we also needed to do something now. As new engineers started trickling in, Samuel and Wes were spending more and more time in their DMs, feeling good about how helpful they were but feeling exhausted by the end of each week.

Wes could see the problem very clearly: there would be no time to invest in onboarding and usability if they continued to be everyone's personal support engineer.

Developing an on-call support rotation was the most immediately obvious approach to fix the growing problem. This would remove the reliance on DMs and spread the burden across the entire team, greatly minimizing daily thrash for everyone but the primary and secondary on-call engineers. However, this attempt at changing the process had its own challenges. Engineers on the DevOps team had already begun to develop specialization and preference for certain systems, and some just frankly weren't as excited about patiently helping those who hadn't RTFMed[2] or tried to figure things out themselves, especially knowing that Samuel and Wes would eagerly take on the responsibility. In addition, Samuel specifically had a hard time letting go of his DM conversations—he had made a lot of friends in those DMs and could feel his impact distinctly. What's more, how do you convince a growing organization of people to follow a new process when that process feels less personal and less efficient (it wasn't actually less efficient, but to those with a direct line to Samuel, it was a real fear). To some of the old guard at Lyft, a new process like this was a foreshadowing of things to come for their beloved scrappy-startup culture. It was an ushering in of new structure, process-for-the-sake-of-process, and red tape. We had our work cut out for us.

When I joined the team, I understood that both the DevOps engineers and those outside the team would see the large, public companies on my resume and view me with immediate suspicion, so I set out to make the smallest change possible in the creation of this new process. Effective immediately, there would be no more support conversations in DMs. All questions and requests for help needed to transition to the team's public Slack channel. Engineers could continue to DM their favorite DevOps member at will, but in response they'd be told, "I'm happy to help you with this; can you please restate your question in the #devops channel so that everyone can benefit from our conversation?" This was the perfect first step because it was entirely within our team's control. Our customers still got the help they needed without any major changes to their workflows, and Samuel, who had

---

2 "Read the f***ing manual": a common euphemism used toward those who ask questions without doing the bare minimum of research first.

to change only the venue for his conversations, got to continue helping his friends and doing the work that he enjoyed.

This miniscule change had several immediate benefits, the first of which was visibility. Suddenly, the rest of the team and I had full visibility into the volume and types of support requests that were being made. I knew just how much time Samuel and Wes were spending on support and debugging. I knew which areas were showing the most strain and which were the most stable. I knew which customer teams tended to run into trickier problems and which needed to develop a bit more self-sufficiency. DevOps engineers themselves gained an appreciation for one another's knowledge and time spent helping others. And now we had problems and solutions stated in a public forum that could be indexed, searched, and referenced in future support conversations. Knowledge could more easily be shared across the DevOps team as well as the larger engineering team.

The second benefit was increased efficiency and scale. With conversations now happening in public, other people could chime in. And it wasn't just the DevOps team monitoring the channel; most engineers at Lyft parked in #devops, and a few would jump in if they'd run into the same issue before. Something Samuel might be able to muddle through would actually fall into Wes's area of expertise, and Wes would take over to provide a quicker and more thorough answer. It also reduced a little bit of strain on DevOps engineers. They could step out in the middle of the day for an appointment and not be so worried about leaving their coworkers blocked, as others could pick up the conversations for them.

The final major benefit was that we were training the rest of engineering to break their old DM habits and get used to a new world of collective responsibility. It only took a few weeks before people started asking their questions directly in the #devops channel first, and this quickly became the default approach as new engineers joined the company every week.

Of course, these benefits didn't last long. Now that answering support requests was everyone's responsibility, it became no one's responsibility. Wes and Samuel were still shouldering the majority of the support burden as they addressed most of the support requests. Questions were slipping through the cracks, especially if they were complicated and/or uninteresting. Other teams would notice when questions went unanswered, and I knew we needed to protect the team's reputation as a customer-focused team. We moved to the next phase of our plan, which was to officially establish a primary and secondary on-call support rotation. Primary on-call's job was to staff the channel, answer questions they could answer, and escalate those they couldn't. Secondary on-call's job was to provide backup to the primary and ensure there were no gaps in coverage during working hours.

This next phase of the process had its own benefits and challenges. It helped to make sure we weren't missing requests (good for our reputation), and it gave everyone on the team equal responsibility over fielding support requests (good for team fairness). Our customers' workflows didn't change at all. Our whole team felt visceral pain when log processing was grinding to a halt, signaling a high-priority investment opportunity that we were motivated to address. However, we were still fighting with individual specialization. We had some team members who were content with the on-call responsibility but others who dreaded it to the point that they'd let questions fester. I had to make cheeky threats to cut off Samuel's Slack access if he didn't stop covering for others on his off weeks. Meanwhile, headcount continued growing; we were onboarding new DevOps members every other week and trying to train them up quickly to join the rotation. The volume of requests was increasing such that the secondary on-call engineer was no longer a backup but instead provided burst capability, until they basically became a second primary on-call.

So we moved on to the next phase, and the next, and the next, adjusting our process to each new challenge. We added in a formal ticketing system that we could integrate with Slack. We added priority tiers and SLAs (service-level agreements). We eventually did split the teams and thus were able to specialize our support rotations. We also did a ton of outreach. With every process change came an all-hands presentation about what we were doing and why. We created time to invest in onboarding presentations and usability improvements, and regular recorded tech talks. We moved DevOps engineers (now rebranded under the umbrella of Infrastructure) to Product teams and product engineers to Infrastructure teams, which helped greatly with the spread of knowledge. We instituted new policies to request that engineers who were new to the company seek help from their immediate team first before escalating to Infra support. Each time, things would stabilize for a bit until the next growth inflection point, when we'd have to readjust our approach again and go back on the campaign trail to explain what was changing and why it would be beneficial to our customers.

The solutions to our problems were not novel. Countless engineering teams who've found themselves in the middle of a growth spike have taken very similar paths. The interesting meat of this story is in the iterative approach. Do a small thing that has more benefits than drawbacks, wait for everything to stabilize, and iterate just as the drawbacks begin to outweigh the benefits again. Look for places in your current process that are likely to become bottlenecks (hint: any part of the process that relies on an individual person will eventually be a bottleneck!), and think about how the process will degrade at scale. Solve for a more graceful degradation.

This period of my career also taught me a lot about introducing process to people. I had to be thoughtful about leveraging Wes's strengths and allowing Samuel to do what made him happy while still optimizing for the health and sustainability of every individual on the team. We had to create right-sized processes that were easy to justify to others and would provide immediate benefit and relief. Wherever possible, we emphasized the benefits for our customers over the benefits to our team, even though the two were inextricably linked. And at the core of this effort was trust. I needed my team to trust me, and I needed the engineering org to trust my team. Placing people's anxieties and needs at the center, and then taking small steps that allowed us to introduce iterative improvements without majorly disrupting our customers' workflows, went a long way toward allowing us to make bigger changes over time. It also gave us space to work toward that medium- to long-term plan in a way that complemented our on-call efforts at every phase of iteration.

BUILDING THE PROCESS LAYER

Process is often a hotbed of debate, and when I think about why that is, I think about penguins. Bear with me, and I'll explain.

Close your eyes and imagine a penguin. Really see the penguin. How big is it? What's it doing? Who are you in relation to the penguin?

Take your time.

...

Done? Keep the penguin in your thoughts. I have an important question.

Do you want to hug the penguin?

When you ask this question, you find a variety of opinions on hugging the penguin. This makes sense, because there are a wide variety of penguins (and also, I'm told some people just don't like penguins).[3] Fairy, or little blue, penguins are a bit over 30 cm tall, and I am not an animal expert, and this is not advice, and y'know, see if the penguin seems open to it, but...totally hug that penguin.

The largest penguin existed in prehistoric times.[4] That penguin was pretty massive, around the size and weight of a human. Now that kind of penguin...you do you, live your best life, but hugging probably shouldn't be your first move there.

As keeper of the penguins, you may think you're offering up a little fairy penguin that absolutely would be a good pet to take home and keep in your bath.[5] And people react like some giant prehistoric penguin from the before-times could very well annihilate them (Figure 12-1).

And you're like...it's just...a penguin?

And they're basically screaming at you: "Arghhh, it's a penguin, oh my god, a penguin, arrrghh!"

Figure 12-1. Hug the penguin

What do penguins have to do with process? Replace the word *penguin* with *process*, and it should start to make sense. When you think you are offering up a small and reasonable process change and someone reacts like you're proposing

3 ?!?
4 Before COBAL. Even before ENIAC!
5 Please note this is a metaphor and not a suggestion.

something horrific, you need to understand why that is and reconcile your understanding. Maybe they are afraid of all process, maybe the process seems bigger to them than you think it is—and that might be true—or maybe you need to make it clearer what the process *actually* is.

The key to adoption of process is reconciling the penguins.

Experimenting (on One Another)

Teams that think they are doing fine can be more resistant to process: things are good the way they are, so what needs to change? What if it makes things worse?

To encourage more openness and flexibility, I use the concept of "experiments," a technique I learned from Emily Webber. "Team experiments" are things that we try to improve the team dynamics or effectiveness. But experiments themselves are a process (we track them, review them, and iterate):

- The concept of "experiments" frames the agreement around process that we will try things and iterate. Not everything has to be perfect.

- The process question is reframed as "What could we try?" This seems much less threatening.

- Experiments support a mindset of continuous improvement and openness to try things. Over time, this mindset shifts the culture of the team.

Let's talk about people

Gretchin Rubin has a fascinating and useful framework for how people respond to expectations called the "Four Tendencies." There's an online quiz you can take (*https://oreil.ly/8cIx9*), but I'll give you a quick overview.

First up, the difference between inner and outer expectations: inner expectations are the expectations you have of yourself, and outer expectations are the things other people expect of you.

The four tendencies are:

Upholder
Readily meets inner and outer expectations

Questioner
Meets only the expectations that make sense to them

Obliger
Readily meets outer expectations only

Rebel
Resists all expectations

So if you've ever wondered about the elaborate strategies some people have for doing things that come very easily to you, you're probably an Upholder. If you get annoyed when people ask you to do things that don't make sense to you, you're probably a Questioner.[6] If you ever feel resentful that you're doing everything everyone else wants and nothing for yourself, you're probably an Obliger.[7] If you just want to be free, you're probably not reading a book chapter about process, but you could be a Rebel.

So why is this relevant? Well, when we're trying to create change—and process is a form of change—we need to get other people to adopt it. Your process is in some ways an *expectation* that you need to give to other people, and this is a helpful framework for how people can—and will—respond to it.

So Obligers will hug most penguins. Upholders are inclined to hug penguins but not if it conflicts with an opportunity to hug a red panda. Questioners will not commit to hugging the penguin until they understand the provenance and motivations of the penguin. Rebels run away before they can be introduced to the penguin.

Or to put it another way: Obligers are pretty easy to give process to (until they rebel). Upholders are a little more critical, but as long as it doesn't conflict with some other expectation, they will probably go with it. Questioners may adopt your process, but you will first have to satisfactorily answer all their questions. And Rebels are almost certainly going to resist your process, but don't take it personally—Rebels resist everything.

---

6 When we did this exercise, my work BFF Eli said to me, "I know you're a Questioner because of how much you love spreadsheets." Turns out, that is one of the criteria in the book for identifying questioners. A total giveaway!

7 Or a woman living in a patriarchal society.

Let's talk about process

What even is *process* anyway? One of the many definitions in *Merriam-Webster* is "to subject to or handle through an established usually routine set of procedures."[8] In other words, process is how we get things done in a consistent way.

When considering organizational change, the three most important aspects of process are:

- Process as an agreement about how we work
- Process as a way to answer questions
- Process as culture

Processes that function as an agreement about how we work help people know what they can expect from one another (or in general). For example, we might agree that we are doing code review in a certain way, using Gitflow, releasing every Tuesday. We might have agreements for what topics get covered in meetings and what can be asynchronous. We might have a process for how we approach taking vacation: what we communicate, where, and when. In large organizations, that agreement is often enforced by more and more arduous process, but in smaller ones, agreements are often lightweight and barely considered process at all. It may be enough to just state the agreement and ask people to adhere to it.

Processes that function as a way to answer questions often center around the documenting and retrieving state. For example, we might commit to certain formats around project updates, use certain tags to indicate the status of an issue, or have a shared calendar for important team events (meetings, releases, vacation). As organizations grow, there's more and more value to being able to consistently answer the questions like "Who is working on this?" "How's this going?" "When might I expect it to be done?" Particularly interesting to me is the process that answers questions that might otherwise not get asked; for example, office hours answer the question "When can I talk to $person?"

Processes that determine the culture often reinforce (or undermine) an organization's stated values. It might be how you structure meetings to make sure everyone speaks—or not. Or how a performance cycle promotes and supports a growth mindset and continuous improvement—or not. Processes that treat emergencies as emergencies and try to reduce them, or processes that reinforce a hero culture.

---

8 *Merriam-Webster*, s.v. "process (v.)" (*https://oreil.ly/WjwE-*), accessed April 9, 2024.

When process meets people

In the end, many processes touch all of these aspects, and these aspects can help create buy-in. Process often gets sold based on the ideal value it creates, and resistance based on perceived overhead, but agreement and adherence typically start with agreeing on the problem to solve.

Real change is both individual and systematic. Process changes systems, but shifts in mindset change individuals. Addressing both together is key to creating meaningful organizational change.

When approaching something that looks like a process problem, consider these questions:

- What is the desired culture?
- Where is the ambiguity?
- How do we want to work together?

Constructing a narrative around identity can appeal to the Rebels, clarifying ambiguity can be a way to connect to the Questioners, and more consistent agreements about how we work together help both Obligers and Upholders, not only because it gives them standards for themselves but also because it helps them know what to expect from other people.

Pitfalls: The performance of process

I think of process as similar to meetings. Going to meetings may be part of your job, but it's not your job to go to meetings! Process is similar: creating process may be part of your job, but it's not your job to create—or enforce—process. It's critical to stay focused on the outcomes and create buy-in. Process should be the least interesting part of the change you create.

It is rarely necessary to reinvent things in engineering teams, but it is important to be selective. Process should never be adopted wholesale, mainly because it won't be adopted wholesale. But often we can look at individual problems and find some standard ways to address them.

When I was learning about workshop facilitation, an "aha" moment for me was learning that the key to making it great is understanding the outcome and keeping that in mind throughout the entire process.[9]

---

9 Credit to Erin Casali for this revelation.

If the team isn't focused on the outcome, then they're just performing a process.

Now I see this phenomenon not just in workshops that I attend or facilitate but *everywhere*. Management is full of mechanics that could easily become process performances: 1:1 meetings. Feedback cycles. Team meetings. Retrospectives. These are not inherently useful activities (and I'm sure we've all been in some of these that felt extremely pointless). They are useful only *in service of some kind of outcome*.

The key is to consider the outcome as you design the process and to pay close attention to the behaviors that emerge as you introduce the process. Behaviors are the link between processes and outcomes—processes encourage behaviors that create outcomes.

Let's take retrospectives as an example:

Process
> At the end of each sprint,[10] the team gathers to look back, discuss, and agree together on what went well and what could be improved.

Behaviors
> The celebration of positives; open discussion among the team around areas for improvement.

Outcome
> The team learns and adjusts together, becoming more effective over time and more able to predict what they can—and can't—accomplish.

If the behaviors we encourage in team meetings are corrosive and micromanage-y, we'll encourage people to hide their struggles behind overwork and excuses. If the behavior we create in 1:1 meetings is to diminish or judge, we'll erode (or never build) trust. If we turn retrospectives into forums for blame, we'll create a corrosive environment on the team.

Process itself is only useful when it serves some kind of outcome, and that must be kept in mind from the beginning and reinforced in the behaviors we encourage (or discourage). As teams discuss process, or leaders introduce it, it's important to talk about what purpose the process is supposed to serve, including the underlying problems or desired outcomes it is supposed to address. Then

---

10 A set period, normally one or two weeks, during which a defined amount of work is expected to be accomplished with a review at the end.

we need to be willing to review and adjust as we see what happens when the theory meets reality—something that is faster and easier with the constant of an outcome in mind.

Developers talk about hating process, but they really love it. When they love it, they call it "culture." Good process is sometimes invisible. When we call it "culture," it's because the behaviors and outcomes are so much bigger than the process itself. When building out tactics, ask yourself: what culture do you need to be successful? Then ask the culture question instead, for example:

How can we build a culture of collaboration?
What would help? Standups, pairing, team calls...

How can we build a culture that values users?
What would help? User research, support feedback, empathy challenges...

How can we build a culture of sustainable delivery?
What would help? Regular releases, retros, attention to defects and metrics
...

These cultural questions will push you to make your processes coherent and supportive of what you want to achieve.

Tactics as a team API

To a certain extent, the tactics layer is the API for your team. For example, how do you know the status of a project and what you need to do next? When there are clear processes that determine what the state of things is, we don't need to have lengthy meetings or chase many people down in order to have a clear agreement. Questions can focus on why that is the state right now, how people feel about the state, and the discrepancies between expectations and reality.

If you're not inclined to get in the weeds, a good API can help you set up a level of observability into your teams such that you know when you need to get more involved. If your default inclination is to get more in the details (or you get described as a "micromanager"), then aligning with the team on the API so that they are clear about what information you need and can trust that you are getting it will help you scale as your team grows.

This also shows up in how we respect (or don't) organizational hierarchy. As a manager of managers, it's tempting to ask individual contributors to do something "small," but it's appropriate to go to their managers and ask them first—they have a depth of context you don't (hopefully, and if not, that's a different problem). When you go straight to the ICs, you risk overwhelming them and undermining their managers.

The fundamental limitation of human APIs in an organization is that they're great for sharing information (most of us have them already, though they're usually undocumented) but they're poor for decisions. And they miss a key point: trust.

If I ask a computer to add two and two, I can feel very confident that this task will be done to my satisfaction. If I feel it's necessary, I can add a unit test to validate it.

But the things we ask of humans are messier, more complicated, and not well defined in mathematics. If I follow the information API and learn that a project is offtrack, and it's a team or lead that I trust, I will start with "How can I help?" If there's no trust there, especially when it's something we need to rely on or this kind of thing has happened too many times before, it's tempting to start with "What on earth is going on?"

There's a comfort for the mathematically inclined in returning to the certainty and understanding of mathematics, to think in systems and optimize for efficiency of communication between them. These things work, up to a point, but they are too static for the messiness of humans and the chaos of growth. And this is why even the perfect process can only do so much.

Next we turn to execution, the final layer.

Driving Execution

Execution is the "do." It's getting sh*t done.

Now, like engineering managers, penguins have a variety of ways to make things go faster. On land, they "toboggan"—slide on their fronts. In the water, they "porpoise"—jump out of the water. Underwater, they can release air from their feathers, which can help them reduce drag and have a short burst of very high speed—useful for escaping predators.

Engineering managers, well...we have process.

There's a manager joke that says, "If managers call people resources, they can call you overhead." This is reality. Management is overhead, and everything we do is in service of people getting stuff done.

Often, leaders reach a level where they don't see their job as managing execution anymore; this is the work of the line manager who reports to (someone who reports to)+ them.[11]

To a certain extent, this is true: at a level of leadership, if you're worried about execution, you're doing things wrong. But as you go up the ladder, there's increasing power to break execution by being thoughtless:

- Saying the wrong thing and causing confusion
- Making an individual feel discouraged
- Focusing on strategy and missing that teams have nothing to do
- Pushing on something that misses the reality of what teams are working on

Never forget that the execution layer is where progress is made. Without the other layers, it's just activity. However, everything else without this means nothing, because nothing happens.

When thinking about execution, there are two main questions:

- Are there people doing things?
- Is it clear what needs to be done?

CASE STUDY: BUILDING STRATEGIC ALIGNMENT TOWARD EXECUTION

Contributed by Jean Hsu, VP of engineering, Productable

Years ago, I was the engineering lead for the cross-functional Publications team at Medium. When I started leading engineering on this team, the entire team was only five (three engineers, a designer, and a PM), and we were able to stay in close communication with one another and make sure priorities were clear. However, as Publications became a larger focus for the company, the team grew to be one of the larger teams at the company and soon grew to 15+ team members. We realized that to stay afloat, we had to divide the team into three subteams, all with lighter technical oversight and product guidance. Execution started to flounder.

It was challenging to just keep track of all the projects in flight. Seemingly straightforward projects that we expected to take a week or so dragged

---

11 Yes, this is a regex joke.

on, sometimes spanning two or more two-week cycles and falling far below the expected ROI. We knew we had to turn things around to justify the size of our team and to get the team back on track toward business goals.

As we tried to debug the dysfunctions of our team, the product lead and I tried a bunch of things, some of which worked, and some of which definitely did not.

We tried keeping closer track of ongoing tasks and asked engineers to break projects down into more granular Asana tasks with time estimates. But we found that the additional overhead of generating those cards as well as a feeling of diminished ownership by engineers made projects drag out even longer.

We made a few structural changes that increased technical oversight across the smaller subteams. An experienced tech lead took charge of these small teams, and more junior engineers were allowed to step into lightweight project lead roles. This gave them more opportunities for growth and learning.

The structural changes put in place gave us a chance to gain some key insights into how the teams were operating. One crucial insight helped get us back on the right track: the realization that cross-functional understanding was low. Product didn't have a great understanding of where the "gotchas" in the codebase were (and where technical debt built up), so they were constantly frustrated when something that sounded simple would take far longer than expected. Engineers often didn't understand the relative importance of projects so would sometimes get sidetracked and spend far too much time on paying down lower-priority technical debt, leaving the higher-priority projects incomplete. And finally, the collaboration between design and engineering often lacked clarity about the level of polish required in designs (should I treat this as a wireframe or build it to pixel perfection?), so engineers would spend unnecessary time building out new components when existing ones would have been just fine. All told, the insights gained from this restructure allowed us to take further actions to improve operations.

The most impactful exercise we did as a team to build a shared cross-functional understanding of priorities was kicking off the next quarter with a prioritization exercise. We documented everything we had done in the past quarter and roughly sized the projects that had been completed based on a points system based on engineering time. Once we had a number (say, 45 points) that represented the amount of work we had done in the previous quarter, we looked toward the next quarter with an exhaustive list of everything that was on the roadmap. We broke the team up into smaller pairs or trios and asked them to do a lightweight investigation and sizing of a few projects and assign it a number size.

We then conducted a several-hours-long team workshop where we assigned cross-functional groups of four to five team members with the same task: take the existing roadmap (which ended up being over 100 points worth of work) and decide what you're going to prioritize this next quarter, given only 45 points to allocate and staff projects.

As teams explained why they chose to prioritize what they did, we were pleased to find that there was a ton of overlap in the decisions people had made. Having put their PM hat on, even for a few hours, engineers started to build a deeper understanding of competing priorities. Larger projects that seemed like cool engineering projects were set aside in favor of high-impact projects that would improve the experience of publishers on the platform.

This prioritization exercise built excitement and alignment on the team and paid dividends for the rest of that quarter and beyond.

Engineers gained a more visceral understanding of the impact of dragging out a small project, which allowed them to make smarter decisions about when to take on versus pay off technical debt and when to ask for help.

Throughout the next quarter, we found that the team stayed in much better alignment. We dealt with far fewer execution surprises, and features shipped much more quickly. Additionally, the team felt a stronger sense of ownership and pride in their work, knowing that we were working together as a well-oiled machine.

CASE STUDY: THE DAILY STANDUP

I once took over a team that was really struggling, and to my dismay, one of the issues that quickly became apparent was that I could not consistently expect people to show up for work. There was a high expectation of autonomy, no real training of managers, and an absence of typical norms or HR processes. There were multiple root issues of this problem, but one of them was that people were often so isolated that it seemed like no one cared if they were making progress or not. If they got discouraged on a problem, they had no one to turn to, and as the days went by, it might just take them longer and longer to open the computer and try again.

One of the most impactful processes we put in was asking people to do a daily standup, in text, at the start of their day:

- This made the agreement that people show up to work each day more concrete—if they didn't post, someone would notice.

- It answered the obvious question of "What are you doing today/what did you do yesterday?" but it also created insight into the question "Is this person stuck?"

- It was the start of shifting the culture to one where people communicated with one another more and helped one another more proactively.

Of course, there was the expected pushback from people who just did not want accountability, but the most interesting pushback came from those who had high intrinsic motivation and high self-esteem. These people typically knew what they wanted to get done on any given day, they did it, and if they needed help, they had people to turn to. To them, this was an unnecessary overhead, and they didn't see the value of it or why it should apply to them. It's reasonable that people who are generally effective are resistant to processes that are designed to highlight—and address—ineffectiveness. Outlining the desired cultural norms *beyond* the process was much more compelling to them than the process itself, and it helped create buy-in. The value of this process was really in being a foundational change in team communication, which allowed many more interesting and impactful changes to come over time.

CASE STUDY: THE NEVER-ENDING ACTIVITY FEED

One project I inherited at Automattic was this all-encompassing tool that was supposed to allow the user fine-grained control over everything that happened on the site and a simple way to undo things. At the core of it was an activity feed showing everything that had happened, with associated actions, and the user could revert or roll back.

Think back to the last time you got your Git commit history in a muddle, and contemplate the complexity of this. Add in the database, separate and yet entwined with the site software, and all the additional complexity there.

Everyone agreed on the end state for this project—the mission. That was easy, because it was going to do everything. At the point where I got involved, this project had been going on for around 18 months with no end in sight. Multiple teams were working on it. It wasn't clear who was responsible for what, and there was a lot of finger-pointing about who wasn't doing what.

Having looked into things, I concluded there was no way out for this project—the failure was at inception in the definition. The way out was to end it in some way (ideally by shipping something) and then make sure that future projects never played out in the same way.

We extracted individual projects from the monstrous project, defined clarity about a desired (and realistic) end state, and gave each project to a team. One team took on the backups and restore (including migrating old users to the new system). Another team took on building the feed. A third team took on some of the possible actions so that they could be wired into the activity feed as ready.

The team stayed largely as-was, aside from the departure of one lead, who was replaced internally. But execution shifted dramatically. With an end state, people could make decisions and work toward them. With realistic goals, developers could see how the work they did day in and day out fit into what we were trying to build. And with clarity about which team was responsible for what as well as clear parameters around the integration points, the blaming of one another disappeared.

The individual components were shipped separately, and a line was drawn under that particular initiative. With some MVP components in place, it was possible to use data and user research to understand the useful parts of the overall dream and prioritize accordingly.

GETTING SH*T DONE

A grandiose dream is not a strategy. Sometimes, that only becomes apparent at the level where the dream meets the teams who need to do the work. When the problem of execution is people's (understandable) inability to understand what they actually need to do, that is relatively straightforward to resolve practically. For some people, it will be a relief. Others will be sad to lose what they had projected onto the dream, and that can create some upset. Although proofs of concepts can be created through experimentation, production software, built by multiple people who need to work together effectively, needs definition.

There's a reason why all these case studies in execution were predominantly issues of clarity and prioritization. I firmly believe that (most) engineers like to ship things and that teams do not stall by default. Yes, some engineers over-prioritize architecture, or "technical debt," and sure, some people just do not work very hard. But when these problems reach the level of the entire team not delivering, we need to look at the systematic, rather than the individual, issues.

As a leader—especially if you belong to an under-indexed group—there's a risk of being seen as great at execution but not great at being strategic. This is a trap: as we talked about in Chapter 6, great and sustained execution *is* strategic, but if people do not recognize it as such, you can fall into a trap of proving it again (and again, and again) on execution and not getting the opportunity to advance past that. Personally, I look back at the times when I took one for the team to "fix" execution, and while I can see the clear benefit to the organization of shipping something that did not look like it was ever otherwise going to ship, I don't think taking these problems were necessarily good moves for my career.[12]

I think this speaks to the core tension of execution. It is critical and yet also table stakes. Minor changes long before I was involved in either of the case studies would have resulted in wildly different—significantly better—outcomes without my involvement. However, those minor early missteps compound over time. Eventually, it becomes a critically important problem, and the work to fix it is *hard*. But that work is inevitably not valued proportionally to that because that work has only generated adequacy at best: that the project worked on for a year (or more) finally crosses the finish line. Still late. Still not quite the features people wanted.

Much of the time, we're in a position to keep execution a nonissue. Maintaining standard engineering management discipline: breaking projects down, communicating clearly (both around priorities and day to day), delivering (continuously). Earlier and more minor interventions in big projects or significant team shifts can keep execution ticking along without drama.

---

12 Unfortunately, for women, saying "no" is often not a good career move either—hello, double bind.

Tactics + Execution = Effective Day-to-Day

Hopefully, as you made your way through this chapter and the previous one, you saw how these ideas about tactics and execution could apply to your teams. Everything needs to fit together, and few teams do all of them equally well; we all have areas where we could exert more attention to create improvement—that will be the focus of Chapter 13.

The hidden challenge is often in what is valued in the organization you're in. Some people only value execution, which is on some level fair enough because it's the metric by which things are judged. But execution is most effective when it fits into a coherent framework, and execution issues are normally a product of lacking in the other layers.

Leaders go to what they are most comfortable with. We've all met—and usually disliked—the person with a new process for every occasion or, at the other end of the scale, the vision person who thinks that will answer every question once they determine it. These are the most visible failure modes at the executive level. These people never seem to see the impact they talk about; they move on after a couple of years, and the next person rolls back half (or more) of the things they fought hardest for while the original person goes and pushes for the same thing in the next organization, creating a never-ending cycle.

In smaller—particularly founder-led—organizations, the tactics are often the most glaringly pressing layer. There can be a huge impact by putting core processes in place, but figuring out that process is needed isn't hard. What is hard is understanding why those things are not there already (whether lack of need, aversion to process, misunderstanding of how humans operate), what is the right scope of those processes, and how to work with the organization as is to make things better.

As an engineering leader, your value to the organization is in the outcomes you drive. Competency in the tactics and execution layers is table stakes.

However, your advancement is typically tied to you demonstrating your ability to determine, communicate, and execute on a strategy that aligns with the organization you're in. Be aware of this; don't fall into the execution trap.

13

Driving Improvement

Engineers often look for definitive answers—it's part of the training. We take the desired output, and we use our knowledge of the way things work to meet that output. We verify it. We claim it is correct.

Of course, this has some limitations. As Donald E. Knuth said, "Beware of bugs in the above code; I have only proved it correct, not tried it."[1] I take this to be about the limits of mathematical justification applied to complex systems.

When we apply organizational reasoning to actual humans, we face the same problem. The constraints of working in a regulated field like medicine or finance are vastly different from the constraints of working for a small startup trying to find product-market fit. The constraints of running a larger, more established team are different from those you face when building out a small team in a smaller organization.

There are many people out there who would like to tell you how to run your team. I am not one of them. I believe the solutions of team functioning are uninteresting compared to the problems and causes of teams that don't function as well as they could. In this section, we will talk about solutions, but to focus exclusively on solutions misses the point. This section is about the conditions that make those solutions useful.

---

1 Donald E. Knuth, "Frequently Asked Questions" (*https://oreil.ly/ejYXE*).

Individuals scale through their own efforts. Teams scale through the structures and clarity that make groups of people work together more effectively. Early on in the leadership journey, new leaders may see their value in knowing things. Their effectiveness in this realm is invariably limited by their ability to argue about and force through solutions, and the limits of that leader become the limits of that team. Sometimes that works, for some period, but great teams do not live within the limits of one (or two) individuals; they transcend them.

The antidote to individual limits is curiosity. When we get curious about teams, curious about the people within them, curious about what is happening and *why*, we create space for alignment and growth. It can be hard to build consensus around a problem, but that is often far easier than forcing adherence to a process designed to address a problem that not everyone believes exists.

In coaching, there is a concept of "powerful questions." Powerful questions are used to get to what truly matters. The questions we will go through here can be powerful questions of team functioning. They are questions not just for yourself but for those around you. If right now you manage ICs and tell yourself that they have no interest in these things, I would ask you if you have tested that assumption. Just because someone says, "I don't want to be a manager" doesn't mean they have no interest in how the team functions or lack the ability to shape the team. As your career progresses and you find yourself managing managers, these are the questions that you can ask them, to help *them* unlock the functioning of their own teams.

Paired with the concept of powerful questions is the idea of "reflective enquiry." This is explored in depth in the book *Coach the Person, Not the Problem* by Marcia Reynolds (Berrett-Koehler), but you can think about it as three steps:

1. Listen and understand what the person is telling you.

2. Reflect back to them the core of what they are saying.

3. Ask them an open question.

A note on open questions: these are exploratory questions with no clear correct answer. For instance, "What process are you going to implement to fix this?" is a closed question that suggests an answer (a process). "What do you think would be most helpful here?" is an open question inviting exploration of what's next.

The four questions don't map to the layers; it's more like a grid (Table 13-1). They can be relevant at every level of team functioning. As we work through them, we will explore the kinds of problems that prompt these questions and the practices that help to solve them; this should make them more tangible and practical in their use. My hope, though, is that you apply the curiosity in your own organization, within your current constraints, to create effective change that works with the unique situation you find yourself in.

Table 13-1. The four questions that map to team functioning, with each applying to the layers of team communication

	Mission	Strategy	Tactics	Execution
How do I create clarity?	Understanding team/company mission	Laying out strategy and measuring progress on it	Team interoperability	Individual progress and accountability
How do I create capacity?	Focus	Staffing	Team productivity	Individual productivity
What am I incentivizing?	Alignment with mission	Alignment on strategy	Team-first mindset	Individual motivation
How are my feedback loops?	Product-market fit	Metrics	Team improvement	Individual growth

To butcher Tolstoy would be to say, "Happy teams are all alike; every unhappy team is unhappy in its own way." But actually, I don't agree with that; unhappy teams tend to be tediously predictable: not delivering, feeling like victims, lost.

When you have a struggling team, the first question is often "Where do I begin? "(Or maybe, "What have I done?"). But I hope these questions are more useful. We often have many viable options for what to begin with; the point is that we do.

By contrast, great teams are self-improving. What does it mean for a team to be self-improving? Self-improving teams have feedback loops that make getting better over time a team effort, they respond well to failure and learn as much from it as possible, and they use estimation as a way to better surface the known—and unknown—unknowns. They invest in collaboration that levels up individuals and the collective. Self-improving doesn't mean self-managing—you, the leader, are not absent in this process. It means that the team members can suggest, own, and drive improvements to how the team operates and what the team delivers.

This reframes the role of a team lead as someone who enables the team, who acts as an accelerant. It means moving away from the need to control, or the feeling that the value you bring comes from the power you have, toward a coaching mindset and the acceptance that the value you bring is shown best in output and team health rather than anything you actively, and obviously, do. Your job becomes the questions: improving clarity, increasing capacity (removing bottlenecks), aligning incentives, and building better feedback loops.

In the previous chapters, we talked about the layers of communication and how they map to aspects of team functioning. In this chapter, we will talk about how to operate on those layers in order to allow the team to decide and drive how they can execute better.

Creating Clarity

Every struggling team I have encountered seems to be experiencing some kind of existential crisis about "who we are" or "what is our purpose," aka the mission question. But as a pragmatist, to me this often seems beside the point. If we're not shipping, how much does it matter what we're not shipping? How can we possibly know what we should be doing two or five years from now if we don't have a consistent idea of what we are doing today?

Here's the thing about the existential crisis: it's comfortable. It's a team-bonding experience because it usually unites them against a common enemy—this is not a good team-building strategy, but you can't fault its short-term effectiveness. No one on the team feels threatened by it because it's largely someone else's problem. Everyone can have an opinion about it because it's abstract enough that most people won't have to do anything to shape or change it. Vision debates are the "bike-shedding" of team purpose and structure.

Clarity is harder because it's more immediate and has to be based on what's happening today (remember that good strategy is based on the current reality), which means you need to confront what's actually happening today—there's no hiding behind "lack of vision."[2] Are teams delivering? Are projects drifting on and on with no end in sight? Are we doing things because they are interesting or because they are valuable?

---

2 Yes, this is another plug for the brilliant book *Good Strategy/Bad Strategy* by Richard Rumelt.

The more concrete we get, the harder it is to get everyone in agreement. A wide range of opinions can find shelter in a broad, fuzzy mission statement in a way that is not possible in a narrow and clear one—the kind that articulates a proximate objective. Clarity involves hard conversations, hard truths, and defining one, then two, then three steps ahead. It is hard work that looks small from the outside—a "no" here, a "no" there, a refinement of this and that. It's often stating what at least some people believed to be true, anyway, such as what isn't going well or what the next steps include.

And hands down, it's the most effective way I've found to stop teams from drifting and get them to start executing.

CREATING CLARITY: MISSION

To create clarity on mission, we need to define and socialize the team mission in the context of the company mission. You'll need to get clarity on how the team fits into the overall organization and make sure you're aligned with key stakeholders, like product managers and related teams. This is an area where managing up to ensure you get direction (as we discussed in Chapter 6) can be extremely helpful.

This may mean:

- Explaining how the team fits in with the broader goals of the organization
- Refining the elevator pitch for the value of the team to the organization
- Defining what the team mission *is*
- Defining what the team mission *isn't*
- Setting some high-level ambitions for the team
- Outlining what the growth of the team is likely to be
- Defining the most important metrics for the team to pay attention to

CREATING CLARITY: STRATEGY

To create clarity on strategy, we must define the strategy and then continue reiterating by reporting progress against it. Again, this work will need to happen in conjunction with those it affects, such as your boss and your peers, and you'll want to seek out input from those people who will need to deliver on it. If you have a product team, they may own the "what" you're delivering while you own the "how" of delivering it, and those things will need to be aligned.

This may mean:

- Articulating the current priorities of the team, the purposes of those priorities, and the rationale behind making them priorities
- Defining the high-level theme(s) of work the team will work around and why those things are important (and clearly align with the mission)
- Defining the key milestones that take the team toward the strategic goal
- Regular reporting on the key projects or themes of work done by the team, why they are important, and how progress is tracking
- Cultivating an understanding for everyone in the team as to how their work fits into the mission

CREATING CLARITY: TACTICS

Clarity around tactics often takes the form of team interoperability: making sure teams know what they can expect from one another and how to get it. Here you'll want to take the time to understand what other teams expect from you and any existing company processes it makes sense to align with.

This may mean:

- Having consistent expectations about standard processes as it makes sense (e.g., project status reporting, PR reviews)
- Having clear instructions for how to get help from the team (e.g., time to PR review)
- Having a defined release schedule and clarity about how to get something in the release
- Having clear staffing: who is working on what and what is the very highest priority (i.e., uninterruptible)
- Determining time frames that you can predict (e.g., focus on the media experience through the end of Q4) and highlighting parallel work that will determine what comes next
- Clearly defining the scope of upcoming milestones in every current (and new) project

CREATING CLARITY: EXECUTION

Clarity around execution focuses around individual progress and accountability. This is where there's most value in co-creating with the team, to make sure you are really understanding their challenges and improve buy-in.

This may mean:

- Ensuring project status is visible
- Adopting daily text-based standups
- Holding sprint planning and retrospective meetings
- Clarifying team expectations
- Ensuring effective and timely PR reviews

THINGS TO CONSIDER

Clarity is important in part because it surfaces disagreement—there is no true agreement until everyone is clear on what they are agreeing to. So when pushing for clarity, don't get discouraged by argument—remember, that is the point. Sometimes you will have to set the clarity, such as when you set the expectations for the individual, and sometimes you will need to facilitate the clarity, such as when you get product, design, and the tech lead all on the same page.

Creating Capacity

Struggling teams also commonly suffer from a sense of overload. Usually, this is very unevenly distributed across the team: some people are exceptionally relaxed, whereas others seem alarmingly overwhelmed. If you view your team as a system, you'll see where you have bottlenecks that, if cleared, would increase the overall capacity of the team.[3]

The goal of addressing capacity is to have the team be able to do more, in a sustainable, equitable way. You can create short-term capacity by forcing people to overwork, at the expense of long-term capacity (when they quit). Increasing capacity *sustainably* involves changing the way you operate such that you can deliver more over time. Sometimes this will involve adding people, but as we discussed in Chapter 7, sometimes you do need to figure out how to do more with less, and this kind of thinking will help you accomplish that.

---

3 Two books I really like on this are *Thinking in Systems: A Primer* by Donella H. Meadows (Chelsea Green Publishing) and *A Beautiful Constraint* by Adam Morgan and Mark Barden (Wiley).

CREATING CAPACITY: MISSION

Creating capacity on the mission is about improving the focus. This largely means deciding what is not going to be done (yet, or ever) and ensuring there's agreement from the relevant people.

This may mean:

- Identifying and cutting work that doesn't align with the current mission
- Redefining the mission to be in-line with what the organization truly needs from the team

CREATING CAPACITY: STRATEGY

Creating capacity around the strategy often starts with staffing: understand what staffing you have and how to deploy that staffing effectively. You might use this to justify a request to add additional staffing, or you might use the staffing you have (or what you can realistically add) to clarify the amount of work you can take on. Again, this will not happen in a vacuum, but also it can usefully inform how you prioritize and spend your time—for instance, if staffing is your biggest bottleneck, then any hiring work will have a higher priority than it might otherwise have had.

This may mean:

- Identifying the strategic importance of different workstreams and allocating staffing accordingly
- Figuring out what you can invest in the "big bets" or longer time frames and what you need to keep the lights on and ship important (but less dramatic work) on shorter time frames, and allocating staffing accordingly
- Figuring out your most critical roles for overall team delivery and prioritizing accordingly in your hiring requests

CREATING CAPACITY: TACTICS

Creating capacity around tactics is about improving team productivity. Here, you'll want to ensure the team understands and buys in to what you're trying to accomplish, and that you understand the reality of their day-to-day work.

This may mean:

- Making process changes for efficiency
- Prioritizing underlying technical work that will reduce ad hoc bug fixing

- Improving developer experience (e.g., release process, build tooling, etc.)
- Building testing infrastructure
- Shifting from external deadlines to team-driven estimations, dates, and ownership
- Adjusting workload allocation to make it more equitable (training people up as necessary)

CREATING CAPACITY: EXECUTION

At an execution level, capacity is about individual productivity.

This may mean:

- Aligning people closer to the work they want to do
- Addressing performance concerns[4]

THINGS TO CONSIDER

If you manage managers, improving them as leaders will always create capacity on your team: their teams will get better, and you can rely on them more and delegate more effectively. Sometimes, it feels like we don't have time to coach the people around us, but as we discussed in Chapter 10, particularly when teams are struggling or growing, in the medium term—not even long term—we don't have time not to.

Aligning Incentives

The work of aligning incentives is the work of making sure everyone is pulling in the same direction. It's making it clear that the work that needs to be done and is most critical is the work most valued by you, and (as much as possible) by the organization more broadly.

In almost every situation, you will get the behavior you incentivize. If you say you value feedback but silence dissent, you won't hear what people are thinking. If you say you value simplicity and user experience but promote people based on complexity, you will get complexity. The gap between claimed values and actual incentives is organizational hypocrisy.

---

4 In a humane way, as discussed in more depth in Chapter 9.

Being clear about incentives and then rewarding the things you want to see are key to making sure teams are aligned. If you want to incentivize collaboration, you can't evaluate purely on individual metrics. If you want to incentivize honesty, you have to be willing to listen to the things you don't want to hear. If you want to deliver user value, you need to make sure you recognize that work. *The role of a manager in a self-improving team is to ensure that incentives align with team improvement.*

ALIGNING INCENTIVES: MISSION AND STRATEGY

To align people with the mission and strategy, you will need to ensure that they understand the mission and strategy and that work is aligned with the mission and strategy.

This may mean:

- Ensuring all work aligns to the mission and is clearly articulated as such

- Eliminating work that doesn't support the mission

- Giving people who don't align with the mission and the strategy options to go elsewhere

ALIGNING INCENTIVES: TACTICS

Alignment of tactics is about building and reinforcing a team-first mindset.

This may mean:

- Rewarding and celebrating the work that makes the team more effective, such as handling releases, onboarding new people, doing effective PR reviews, and so on

- Making it clear that work that advances the team is more important than work that advances the individual (i.e., don't put off or refuse PR reviews for the sake of individual productivity)

ALIGNING INCENTIVES: EXECUTION

Aligning incentives at an execution level is about individual motivation.

This may mean:

- Ensuring that individuals understand how their work fits into the overall goals of the team

- Building a narrative around the most impactful projects to make them considered promotable (i.e., make the impact clear, surface the real complexity, and avoid incentivizing unwanted complexity)
- Making the impact of the (nonobvious) work that supports the mission clear (e.g., infrastructure, bug fixing, etc.)

THINGS TO CONSIDER

When incentives are misaligned, at the extreme you see teams or individuals working against one another, creating confusion, disruption, and waste. But the less extreme ways can also make a team less effective, particularly in terms of tactics, when individuals prioritize their own effectiveness over the collective impact of the team.

If clarity is where you agree on the mission and strategy, and capacity is where you make it possible, aligning incentives is where that agreement is made real—and reinforced.

Building Feedback Loops

People only improve consistently if they get feedback, so feedback loops should be (1) tight and (2) positive—reinforcement of what goes right. This doesn't just mean feedback on what they can do better, although that is important. Feedback is just someone's work reflected back to them, and making sure people get consistent feedback and appreciation (an incentive!) for what they are *already doing well* is crucial. Giving people only constructive (or worse, negative) feedback will cause them to avoid you.

Say we launch a new feature—we want to set up feedback loops around if anything goes wrong, like monitoring for downtime, support volume, and so on. But it's also important to set up the feedback loops for what we expect to go *right*, like usage, positive feedback, and the like. *The role of a manager in a self-improving team is to build and contribute to healthy, productive feedback cycles.*

BUILDING FEEDBACK LOOPS: MISSION AND STRATEGY

Feedback loops on the mission and strategy show your progress against them. For instance, if your mission is around product-market fit, the feedback loops show your progress against that.

Opportunities for feedback loops include:

- Highlighting customer feedback (including internal customers where applicable)
- Highlighting wins in the market (e.g., hitting a new price point or opening up a new market segment)
- Reporting usage metrics that tie to strategic priorities
- Reporting execution progress against the overall strategy

BUILDING FEEDBACK LOOPS: TACTICS

Feedback loops around tactics are often around team improvement.
Opportunities for feedback loops include:

- Team health metrics (e.g., engagement surveys)
- Sprint feedback
- Velocity metrics
- Defect rate
- Celebrations of team wins

BUILDING FEEDBACK LOOPS: EXECUTION

Individual feedback loops are centered around individual growth.
Opportunities for feedback loops include:

- Project impact
- Retrospectives/lessons learned
- Peer feedback
- Celebrations of individual wins

THINGS TO CONSIDER

Feedback loops are a critical tool for motivation, alignment, and reinforcement. They can help people feel progress during a long project and build confidence that a strategy is working—or at least let you know sooner if it is not.

Where to Begin

Every new manager struggles to let go of the individual work they were doing before and had achieved some level—presumably—of mastery at. It's easy to claim this ideology but hard to truly enter into this mindset of team facilitator. We talked about the failures of the idea of servant leadership in Chapter 5, and managers can also use the idea of team responsibility as a reason to abdicate responsibility for the day-to-day of their team, as if responsibility can only be a product of control (although often, not even then—and power without responsibility is a form of sociopathy).

In Chapter 7, we talked about how the best leaders master multiple leadership styles and know when they are best deployed: when to take the hard line of the authoritative stance, when a situation is bad enough that a coercive style is necessary, when a team would benefit from someone setting the pace. Mastering these different styles is knowing when to optimize for the increased buy-in of the democratic approach and accept the slower progress in the short term, when the team needs an affiliative leader to help them bond, when showing up as a collaborator will get the best out of people, and when someone needs a coach.

The self-improving team is the place for the democratic and affiliative styles but, above all, the manager as coach (and collaborator). Like all good coaches, you will want to ask good questions. Like any good collaborator, you need to be willing to get in there and provide active help (as discussed in Chapter 9).

It's likely that asking these questions has given you a whole lot of work to do. As you are human and can only do so much at once, now comes the question "What do I do first?"

Answering this question is a balance between impact and time. The highest-impact things you can do may take a while, but there are usually several relatively impactful things that won't take too much time. If you're new to leadership or the team, you'll also have to balance building trust and demonstrating impact. For example, clearly communicating big-picture information about what is going on to the team is a worthwhile exercise that often uncovers new information, is usually relatively quick, and creates shared understanding across the team of where projects are at.

However, these shorter time investments will not have any impact unless we also take on some bigger ones. For example, there is nothing more corrosive to overall output than if one or more teams have a lead who is struggling or just isn't a good fit. Addressing this kind of situation is never straightforward and never easy (as we covered in Chapter 9)—the only thing worse and harder, ultimately, is letting it continue unaddressed.

Every situation is different, so my suggestion is to write down a list of ideas for each; talk them over with your boss, a trusted peer, or a coach; and then start taking them on one or two at a time. Identify your key collaborators and build relationships with them such that you can work on things together—or align and divide. Just remember that the list isn't static: as teams evolve, the bottlenecks change, and you'll need to keep asking these questions, revisiting and adjusting as you go.

Your Action Plan for Making Your Team Self-Improving

We've covered a lot in this section, so here's a suggested path for helping your team become self-improving. This is not a one-time checklist but a mindset to be developed and reinforced, and getting the team to adopt it will take time. This list is long, and maybe a little overwhelming, but you're not supposed to go through it all at once—review your next action and give yourself some time to take it, and then come back to it again later.

Step 1: Identify the trend.
Consider how much your team has improved over the last 6–12 months.

1. If the team is regressing or stagnant, it's time for serious focus (start at step 2).

2. If the team is OK but could be better, how can you increase the pace of improvement? Start at step 3.

3. If the progress is good, can you get more people involved and begin driving change? Start at step 4.

Step 2: Identify critical bottlenecks.
Are there any critical issues that need to be addressed?

1. Identify any critical areas of clarity: what is the team's purpose?

2. Identify any critical areas of capacity: what bottlenecks need to be immediately solved?

Step 3: Assess what's working—and what could be better—in driving change.

Consider recent changes you've made—what layer were they in? Did they have the impact you expected? If not, can anything in this section help you revisit?

1. Execution is table stakes, so start here: what could be better?

2. What tactics do you need to support better execution?

    a. Assess your ability to sell and drive adoption of process—do you need to make your penguins more huggable?

    b. Align on the problem before selling a solution.

3. Does your team have a clear mission and strategy?

Step 4: Build your matrix.

Put together your layers/powerful questions matrix and identify highest-leverage improvements.

1. Review your list with your boss, coach, or trusted colleague; get them to challenge your thinking.

2. Build your allies: the people inside and outside the team whose support will help you make change.

Step 5: Make it a team effort.

Foster a culture of improvement on the team, empowering and recognizing the people who are driving the team to be better.

Section 4 Summary

Congratulations! You're now responsible for team facilitation...but, actually, that was always true. Hopefully you now have some new tools and ideas to support you in this.

A good team has clear and coherent communication; this is a product of psychological safety, as is the ability to treat failure as a learning opportunity. They have a good idea of what can be done and how; this makes them able to ship regularly and predictably. They understand why things should be done—and why they shouldn't, so decisions either get made or are escalated appropriately. They have aligned incentives and strong feedback cycles, which allows developing of individuals within the team, creating growth and opportunity.

In Chapter 11, "Mission and Strategy", we talked about what makes for an effective team mission and how to develop a strategy that connects to it. We talked about why these things are important and used case studies to illustrate what that looks like in practice.

In Chapter 12, "Tactics and Execution", we talked about the layers of team functioning, focusing on the tactics and execution layers, which are table stakes for team effectiveness. We looked at how to connect those to the mission and strategy levels, how to create adoption of process, and how to debug team execution. We used case studies to illustrate what that looks like in practice.

In Chapter 13, "Driving Improvement", we talked about the four powerful questions that can be used to unlock team improvement: How can I create capacity? How can I create clarity? What am I incentivizing? How are my feedback loops? We applied these questions to the different layers of team functioning and explained how they can be used to build a mindset on a team that supports the team becoming *self-improving*.

This is the end of the second half of the book, where we focused on teams and making them effective. In Chapter 14, we will outline what a high-functioning team looks like and, finally, try to answer the all-important question of how you know if you're doing a good job.

PART III

Conclusion

14

What Good Looks Like

Over the course of this book, we've been on a journey to figure out what your role is and how to do it effectively. We covered what it looks like to DRI your career, how to practice effective self-management, how to build a healthy team ecosystem, and how to get teams delivering and self-improving.

You'd be forgiven for thinking that "good" looks like doing all of these things well, all the time.

It doesn't. I'm not even sure if that's possible.

So if "good" doesn't look like doing everything we discussed perfectly, and it doesn't look like everyone being happy 100% of the time, what does it look like? And wait, doesn't the concept of an end state contradict the idea that good teams continuously improve?

Good isn't static, and it isn't executing perfection every day. Good is *sustainable*. Good is a situation you could hand over to someone else without guilt. Good is having sufficient breathing room such that you can figure out what's next for you. Good is an opportunity to move forward or upward, or just to take a breath.

Good is a team that changes people. It makes them more in and of themselves; it makes them expect more from the places they work subsequently. Good is a team that people will remember being on for the rest of their careers. Good is a team that delivers business value as a matter of course, and growth and learning alongside because it's the way they want to operate.

This is a chapter about what "good" looks like.

What Good Looks Like in a Team

Inspired by Maslow's hierarchy of needs, this is Huston's pyramid of team functioning (Figure 14-1).[1] Each layer builds on the next, so let's start at the base and work upward.

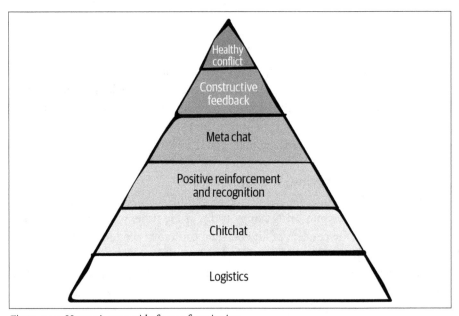

Figure 14-1. Huston's pyramid of team functioning

LOGISTICS

What is it?

Logistics is the information people need to exchange to keep things running: What time is the meeting? Who is running the release? Who is working on the bug?

Why is it important?

Without a functioning logistics conversation, teams are not teams at all—just a group of people pushing code to the same repo. Logistics is what gives people clarity about what's going on and who is working on what.

---

1 Not the snappiest of names, but then naming is one of the hardest problems in computer science.

CHITCHAT

What is it?

Watercooler conversation, exchanging snippets of information.

Why is it important?

This isn't about everyone being friends. This is about people exchanging more information than is strictly needed and having some understanding and care for one another as humans.

POSITIVE REINFORCEMENT AND RECOGNITION

What is it?

Telling people when they have done something good and genuinely celebrating wins.

Why is it important?

This builds trust and fosters a collective mindset rather than an individualistic one.

META CHAT

What is it?

Meta chat is chat about the way we work and how we can be better and more effective.

Why is it important?

This signals investment in the way things get done and people feeling empowered to suggest ways to improve it.

CONSTRUCTIVE FEEDBACK

What is it?

Communicating what hasn't been effective and what could be better, both collectively and directly to other individuals.

Why is it important?

When people trust one another, they communicate directly and don't need intermediaries to communicate things that are more difficult.

HEALTHY CONFLICT

What is it?
Disagreement!

Why is it important?
Healthy teams are not necessarily harmonious. They disagree constructively and resolve that disagreement effectively. This ability to disagree, argue through it, and move forward is a product of psychological safety. It helps make sure more (ideally, all) opinions are shared and helps people reach better conclusions.

Hopefully, it's clear how these different attributes build up to a high-functioning team. At levels below positive reinforcement and recognition, the team is in crisis and functionally not a team at all. With positive reinforcement and meta chat, the team is starting to function and evolve. Adding constructive feedback makes a team good, and a team with healthy conflict is *great*.

This doesn't strictly map to Tuckman's model (*https://oreil.ly/h3v9W*) of "forming, storming, norming, performing," although I believe they make sense together. Roughly, the forming portion corresponds to logistics, norming to meta chat, and performing to constructive feedback and healthy conflict. Storming—which was only observed in about 50% of teams in the original research—is the noise along the way.

What Good Looks Like in Leadership

Something I struggled with as a new manager was finding a sense of accomplishment, and once I reached the point of managing other people in leadership roles, like other managers and staff engineers, I saw this become a challenge for them, too. It's hard to find the right success metrics on which to judge our work because our output is to make the team better, and hopefully, we generously give credit to them.

Without success metrics beyond the team's improvement, though, it can be easy to feel like you're just riding a wave of good people doing good work without contributing anything yourself.

In Section 2, "Self-Management", we covered some of the things that can become success metrics without being useful, like trying to be too helpful or seeing your success metric as being available to your teams 24-7 (which is, of course, unsustainable). Or by embracing the performance of management without understanding the underlying motivations, measuring activity rather than impact. And then there are the obvious ones, like counting lines of code, something that thankfully, as an industry, we have mostly agreed is absurd.

But still, leadership can be a hard job, and you need to have some idea of what it means to be successful. So to that end, I have compiled some suggestions.

CAN YOU TAKE A WEEK OFF?

Taking a week off can be a rough one to start with if you lean toward constant availability as your metric, but there's nothing like a week off (or more!) to show which of your activities has the most impact. When you come back, pay attention to what you find. What's surprising to you? What comes up in your 1:1 meetings? What did people miss? What did they not need you for? As my friend Jill says, "Aim to be unnecessary, but missed."

If your team is in a tough spot, and you don't feel you can really disconnect, try designating one trusted person to check in with each day. Ideally, you should be able to leave some way to contact you in an emergency and then disconnect, confident that if you were truly necessary you would know (and be relieved to find you're not).

CAN PROBLEMS BE HANDLED WITHOUT YOU?

The best example I have of this is the time that one of my team leads was away when a problem arose that required an unplanned release. His team, which was relatively new to working together, found some gaps in the documentation, got wildly creative, fixed it, pushed a new version, and put up a detailed account of what had gone wrong, along with next steps.[2] At some point, both the team lead and I checked in to see if there was anything we could do, but the team had everything handled.

This is huge—you'll never get away from using constant availability as your metric if every emergency must come to you before any action is taken. Ensuring that everyone on the team feels a sense of responsibility and ownership and having a clear directly responsible individual (DRI) for critical things are key. You will also need to have the discipline and self-restraint to stay out of the problem; otherwise, people will never get the opportunity to show that they can do things without your hovering over them. Leave the laptop behind when you go away or, if that's not possible, try turning it off—adding a bit more friction before you "just check in on one thing."

---

2 After the incident, we put things in place such that no one needed to get *quite* that creative again.

DOES YOUR TEAM DELIVER CONSISTENTLY?

Delivery is a trailing indicator for a healthy team, but it is an indicator. Healthy teams ship consistently and keep shipping over time. We all have projects that become unexpectedly complex, and every individual one may have a reasonable explanation, but if you look at the overall picture, is the team delivering more often than not?

DO PEOPLE TELL YOU WHAT THEY THINK?

One thing that we all have to get used to in leadership is people being less candid with us. The higher in the org chart we rise, the less forthcoming ICs are with feedback (even if there's always that one vocal team member). We need to make ourselves available explicitly to people who don't want to presume to seek us out—these are important people to listen to; otherwise, you just hear the loudest voices. Yes, there are still people who lean toward speaking their minds, but if we can create space and listen, we can get others to be open with us, too.

It's also important to note how people give you critical feedback. Do they wait until it's something they are really frustrated by? Or is it an ongoing conversation? Will people give you implicit feedback? Will they tell you what they are worried or insecure about? Will they share what they notice is going on around you? Have you set up a model where you've told people to only come to you with a solution, not just a problem? It's important to foster an open line of communication.

One of my peers once told me, "Cate, I would never let you do anything that stupid."[3] I laughed, of course, but I was also deeply grateful to know that there are people around who will call me out on bad decisions.

DO PEOPLE ON THE TEAM TREAT ONE ANOTHER WELL?

Effective teams are inclusive teams, as we covered in Section 3, "Scaling Teams". Fundamentally, I believe that inclusion is the right thing to do. However, if we consider problems of exclusion—racism, sexism, ableism, ageism—they come from the idea that some people can do better if they push others down, and of course they start with the most marginalized among us.

As a leader, it's on you to cultivate a respectful environment on your team and to make it clear that you will not tolerate discriminatory words or behavior. This is the minimum. Beyond that, you can set some values around reward and

---

3 My work BFF Eli Budelli.

advancement that make it clear that success on your team is something that happens interdependently, not as a competition.

IS THE TEAM SELF-IMPROVING?

As we discussed in Section 4, "Self-Improving Teams", self-improving teams critique and iterate and change things as a part of their process. They're not afraid to discuss what worked and what didn't, make suggestions, and try changes, knowing that some of the changes they make will fail.

These teams get better over time with less and less intervention from you. It can be really hard to get teams reflecting on what went right and wrong with a project because this process is scary (and the first few times might be quite rough). But getting to a place where these retrospectives are a matter of course is core to the mindset of a self-improving team.

CAN YOU GIVE PEOPLE WHO REPORT TO YOU MEANINGFUL, IN-DEPTH FEEDBACK?

As we talked about in detail in Chapter 3, feedback is someone's work reflected back to them, hopefully in a way that helps them take pride in their accomplishments and makes actionable the places where they can improve.

This means having enough insight into their work, accomplishments, and struggles to be able to do that. A lot of that feedback happens as we go, but it's important to regularly (whether or not you have a review process) get some (qualitative or quantitative) feedback from team members and use it to put together a bigger picture of how someone is doing. Think of it as like having a hand mirror in 1:1 meetings and a full-length mirror in a feedback round.

WHAT KINDS OF THINGS CAN YOU DELEGATE?

Do you feel like you can hand off pieces of work or problems to people on your team? Are those projects getting bigger over time? Maybe you started by giving people tasks, but over time, you want to be able to give them broader problems to own. This allows you to take on more from *your* boss.

If you manage managers and don't have people you can hand stuff off to, you will drown. It's just not possible to operate effectively at that scale without the shock absorption of people being able to take things off your plate and handle them. If you don't have it, you will need to build it because it will only get worse over time. This starts with expanding your leadership styles such that you no longer feel the need to control things (Chapter 7) and building a bench (Chapter 10).

WHO IS TAKING ON BIGGER ROLES?

The most gratifying part of my time in management is looking at the people I have managed who have stepped up to be a tech lead, a manager, a staff engineer, or a director, or who, even without a change in role, are generally taking on things with larger scope. As a team grows, there's more opportunity—and more need—for people to step up. Delegation flows down: pushing things onto the managers forces them to push things onto people on their teams, and this is how we grow new leaders.

As much as we may value everyone on our team and want to keep them together, having a strong team means that sometimes people's best path for success lies outside of that specific team. As we discussed in Section 3, "Scaling Teams", it's our job as leaders to help them toward growth and to help our peers when they need a skill set that someone on our team can best provide. It's a sign of success when people from our teams move to different teams and take on more responsibility. It can also be a sign of success when people leave and take on bigger roles elsewhere.

Hopefully, this is something you can talk about together, but the success metric is: how have you helped them? What feedback, what projects, and what responsibilities have you given them? How have you used your insight into their capabilities to help them find their next role inside your organization? If it's more indirect, how has the way you've run the team contributed to creating opportunities for a broader set of people?

CAN YOU TAKE ON WORK OUTSIDE OF YOUR IMMEDIATE SCOPE?

Having our own teams in order and strong support within them makes it possible for us to provide more support to those above and around us.

What could you take on that would most help your boss? Your peers? What scope of things could you take on? Can that get bigger over time?

In both my current and my previous roles, I joined to scale up the mobile team. Once that had grown into a functioning native applications org, with a competent layer of leadership underneath me, it became time for *me* to take a bigger role in the organization, to do more to support other teams, to spread good practices and be part of more conversations.

DO YOUR PEERS VALUE YOUR PERSPECTIVE AND COME TO YOU FOR ADVICE?

Every organization has its own unique set of quirks, and the people who best understand the stress under which we operate are our peers. Who respects who in a peer group says a lot. Pay attention to the topics people seem to value your opinion on. It shows what they notice—which are often the things we most take for granted.

Does Your Team Need to Rebrand?

Even if your team has reached a level of "good" internally, it may take time for the rest of the organization to recognize that. In that case, you may want to think about deliberately undertaking a "rebranding" exercise for the team.

Some leaders are fantastic at "team branding"—communicating about their group in ways that give the rest of the organization a good understanding of what the team is all about. Others are squarely focused on "team public relations"— telling a great story about a team that, if we looked more closely, we might find is not delivering or functioning very well (but it always comes out in the end).

Team PR is usually overly positive, glossing over the hard questions. It's about generating the right illusion. But when what you say (your PR) doesn't align with what people say about you (your branding), that's a surefire way to undermine trust in your leadership.

Don't get me wrong. PR is a legitimate part of the team-branding process. You need to be able to talk about your team's strengths and accomplishments. But to build a brand that reflects and projects reality, you also need to be able to talk about your team's failures and the gap between where the team is now and where it hopes to be.

When I joined the mobile group at Automattic, the team was pretty siloed and generally maligned across the rest of the organization. Laboring under (and resenting) the perception that they weren't delivering, the team was desperate to rebrand. We quickly set out to revive the team's reputation through a PR campaign of sorts. We talked more openly about what we were doing and why, and we became more deliberate in sharing our launches.

The PR effort worked, somewhat, but my boss called me on it and asked what was really happening in the places that I couldn't shine a light on and celebrate, yet. He was right; there were still real issues on the team. Over time, as those issues were resolved and the team became more effective, we could genuinely do team branding.

As I moved to lead another group, I saw that this new team was generally talked about as a product team, but a significant portion of their work was infrastructure. Again, we employed a strategy to rebrand the team so that their critical contributions to infrastructure were better understood. This recognition helped us talk about our work in a way that made the team's impact clearer, which in turn made people more motivated and helped us deliver more.

A well-run, consistently delivering team reflects well on everyone involved, especially those who lead it. But remember to separate the team from your personal brand, and the need for team recognition from your own personal need for recognition.

At the same time, it's important not to go too far the other way. Even if you haven't mastered the art of talking about your own work in a way that doesn't feel like bragging, you need to learn to talk about the work of your team in a way that reflects well on them. Just make sure you learn to do so in a way that is authentic.

Here are some substeps of team PR that you may want to take.

REBRANDING PROJECTS

Engineering projects often focus on the technical details and challenges but always, always need to be talked about outside the team in terms of customer impact. Folks outside your team may not understand the technical details, but they will understand the bigger picture. Even those who do (or think they do!) understand the technical details may not see how those things connect to the higher-level goals. What is the expected impact? Is it addressing common support issues? Improving new user experience? Existing user experience? Better supporting internal customers in some specific way—for example, in release cadence or uptime? What metrics do you expect to affect? Support volume? Signups? Retention?

OWNING ACHIEVEMENTS

It's important to make sure you celebrate and let other people know when you deliver! Within that, you can mark smaller things, sharing individual achievements or things people are grateful to their teammates for.

MARKING PROGRESS

Big projects tend to start and end with a bang. There is the excitement of the initial forming of the team and, eventually, the thrill of the big launch. In between, it's easy for things to turn into a slog, which can lead to a lack of engagement, a spike in turnover, and ultimately, a failure to meet goals.

Remember: all progress is made incrementally, whether or not that's how it's ultimately described. Learning how to talk about progress—and show progress—in ways that people can actually connect with can be a really important part of team branding, especially around execution. Laying out the end state is key, but so is clearly demarcating the milestones on the way and being transparent about progress marked against them. This also can be a helpful way of surfacing and addressing problems early, rather than finding a nasty surprise at the end.

CLEAR RETROSPECTIVES

How can acknowledging failure possibly help with branding? Well, failure is inevitable (the goal can only be to fail less and fail better), so the best way to diffuse whispers of failure is to be open about it. This gives you the opportunity to own the narrative.

When people speculate, they are always missing information. When you are called to answer for mistakes, it's easy to come across as defensive. But when you proactively lay out the circumstances that led to the situation, clearly articulate how it was handled, and describe any process changes that resulted, you demonstrate ownership, accountability, and awareness. It inspires confidence.

It's also an opportunity to thank people outside the team who stepped in. It's rare for the effects of a failure to be isolated to one team. There may have been teams that were depending on you and felt let down, teams running systems that you inadvertently overloaded, teams handling customers that had a sudden surge in requests for support. When others know that you take the impact on them seriously and work to address it, they are much less likely to complain.

REBRANDING PEOPLE

Sometimes, part of rebranding a team means rebranding people—especially those who were seen as contributing to the team's previous problems or who had struggled to be effective under the previous structure.

Sometimes the reputation is deserved, but far from always. Most people don't want to fail. The challenge of rebranding is to understand the reasons for the failure—perhaps some were structural and some were personal—and working to address them. Were people in the wrong roles? Had they not gotten the right feedback? Were they not set up to succeed?

Once you feel like the structure is addressed, you can start positioning people more positively, working with and showcasing their strengths. Often, you also need to rebrand people in their own eyes: helping them see themselves as a leader when they used to be an individual contributor or helping them understand that their strengths lie in technical leadership, not people management. You may also need to rebrand the environment to them, moving them out of a victim mindset into one that is more empowered.

In every struggling team I've encountered, there were people who weren't a good fit and needed to leave to move forward. But also in every situation, there were far more people who, with the right feedback, coaching, and encouragement, were able to surprise everyone (not least themselves) with what they were capable of. Making sure those people receive recognition and credit for their work, in terms of both personal growth and business impact, is a key part of a good team rebrand.

STARTING AT THE END

If the team's brand is what people say about your team, then when you set out to talk about your team yourself, a great place to start is with the objective strengths of the group and what the desired brand of the team is. The objective strengths are the things that are true today that you can already talk about. The desired brand is the gap between where you are today and where you want to get to.

As you consider your aspirations for the team brand, you'll want to put processes in place to support your desired outcome. If you want the team to be known for dependable delivery, you'll need to be sure it stays on target, on top of communicating a lot about delivery. If you want people to think your team has a supportive, open environment, you'll need to do a lot of work to foster that kind of environment, both before and after you talk about it.

The metaphor I like to use to explain the process of branding is the Yoko Ono creation "Grow Love with Me": the word *love* is engraved on a bean, and as the plant grows, the word *love* appears on a leaf. This is brand. The core is expanded, and what is inside becomes what is outside.

When PR is inauthentic, when the outside doesn't match the inside, it's an exercise in fantasy. But when you can talk about your team as is—the successes it leans on, the failures it learns from, the aspirations it's working toward—you've made PR a valuable piece of your team branding.

What Now?

At the start of the book, I talked about how when I moved into a leadership position, I had a much clearer idea of what *not* to do than what *to* do. I had more clarity about what bad teams looked like than what good ones could be.

I realized that to be a good leader, I needed to build that model of what good looked like for myself and instill it in as many people as possible. Writing down that model and publishing it in a book brings it to the widest audience I could have imagined. Thank you for reading. I hope it has been helpful to you.

I started writing this book during the boom times of tech, as the pandemic wound down. We were talking more about employee well-being and strategies for retention than ever.

And then...interest rates went up. Layoff after layoff was announced. Many companies that had embraced remote work once again forced people to return to offices—maybe because they believed the office is actually more productive but probably at least some for the attrition those return-to-office mandates caused. The richest man in the world bought Twitter, and tech people seemed divided by those who thought his approach was horrifying, short-sighted and wrong and those who thought it was "visionary leadership" and efficiency.

In many ways, it is an entirely different world. And yet I think, beneath the noise, some things remain constant.

Teams exist for a purpose, and the role of a leader is to guide the team toward that purpose.

Beyond the purpose, 40 hours a week is a lot of your life. Great leaders can do more than get the team to achieve its purpose. They can create environments where people thrive—they learn and grow *while* they deliver. Where the purpose of the team is just part of the value it creates.

In the boom times, I think we sometimes forgot the purpose. Which is perhaps understandable, when companies would hire people and pay them absurd amounts of money just to keep them from working somewhere else. When we talked about org size like it was the metric that mattered, more so than what the org produced.

But equally, in the dark times, I see people forgetting the purpose. Is it a win to slash all your costs if you've also decimated your major source of revenue? Surely that can't really be the case? "Be profitable" is no more guidance than "grow faster"—neither of these is a goal until it has a strategy behind it. Without a strategy, they are just dreams.

All of this is to say, if the present seems like a hard time to be a leader, that's probably correct. But that just makes all of this more important. Now that it's clear that job security is a myth, it's clearly vital that you truly DRI your career in order to achieve *career* security. As leaders everywhere become more pressured and stressed, the more important self-management becomes. If you have to figure out how to do more with less, then it will be critically important to make sure the team you *do* have functions as well as it can. If you are under increasing pressure to deliver, the more you will need to drive clarity and ensure that you can, in fact, deliver.

You can do it, though. I believe in you.

Team Strategy

Table A-1. *Effects of changes in people and scope*

		People		
		↓	=	↑
Scope	↓	Deprecation	Reorganization	Consolidation
	=	Efficiencies	Stability	Growth
	↑	Layoffs	Expansion	High growth!

↓ People && ↓ Scope: Deprecation

Deprecation happens when something has reached end of life. This can be unmitigated bad news (such as the company is failing) but not necessarily. End of life might just mean a changed scope, which can sometimes be a positive. For example, it's not bad news in the case where the team just owned too many things and needs to be broken up for clarity and improved effectiveness.

For instance, say a team is managing a legacy service and the plan is to migrate to a new one; as such, the scope of what the team owns and of the team itself will shrink over time. Maybe it will go to zero eventually, or maybe one team was supporting disparate things and is being right-sized.

WHAT TO CONSIDER

You'll need to think about how to situate, manage, and communicate the change within the broader context of the organization. What practical things—like knowledge transfer and SLAs—will need to take place? How do you make sure there's clarity on who owns what and that the people who remain are set up for success?

WHAT TO WORRY ABOUT

If it is objectively bad news, clearly you will need to worry about that. However, it doesn't necessarily matter whether or not this is objectively bad news: when teams get smaller, it often *feels* like bad news.

People will worry about their workload, that what they are working on is not that important, and what that means for their career. When teams get very small, it's easy to get to a place where only one person knows certain things, and if they leave, it will make things very difficult for those who remain. It's also possible that you get to a place where the responsibilities on the org chart don't match the responsibilities in reality—for example, there is a new team that is responsible for something, but the old team gets pulled into every incident (but not every code review, which means there are many incidents).

↓People && =Scope: Efficiencies

This situation arises when there's a need to "do more with less." It can be the product of cost cutting, or it can be the result of needing to focus efforts elsewhere—for instance, on new products or addressing some legacy issue.

WHAT TO CONSIDER

As the saying goes, "There's nothing more permanent than a temporary solution," so even if it's supposed to be short term, you want to avoid racking up debt in terms of backlog or temporary, hacky solutions.

WHAT TO WORRY ABOUT

Whether this is an output issue or an efficiency issue—it's tempting to "do more with less" by getting individuals to just...do more work, which, if it lasts too long, will cause morale and retention issues. Good, clear prioritization is key. If you can't do everything you used to, you need to be trying to cut the things that matter less to ensure that the things that matter most still get done.

↓People && ↑Scope: Layoffs

See Chapter 7.

=People && ↓Scope: Reorganization

Over time, as things change, people join, people leave, people follow their interests, and we can end up with an organization that isn't structured to do the things that it needs to do. Change is required to clarify the scope of teams and ensure that those teams are staffed to be successful. You'll find that at the most extreme end of this (such as addressing a historical understaffing and moving people around to different teams), teams can have their scope reduced while increasing in size.

WHAT TO CONSIDER

In situations where there's a dramatic shift, it's always worth considering how things got to that point and, once the obvious changes have been made, what more will be required.

WHAT TO WORRY ABOUT

Even when change is desperately needed, it can still be scary. This can be a significant shift for people on the team and require different ways of working. There's more about change management in Chapter 8.

=People && =Scope: Stability

See Chapter 7.

=People && ↑Scope: Expansion

Expansion is the other way to do "more with less"—using the same people, with increased scope. Whether this feels stressful or like an exciting opportunity depends on where you're at and where you're going. If everything is under control and you have good processes in place, even with a bit of slack, it can be exciting to have the pressure of something new. If you were already spread thin, expansion can tip the balance into becoming overwhelmed.

WHAT TO CONSIDER

Normally, this kind of change is a good one, so consider how to sell it accordingly. The team has been doing well and now gets a new, additional responsibility. It's exciting! An opportunity!

WHAT TO WORRY ABOUT

Expanding in scope can break things that were previously working. If you have a team split between multiple things, they can feel less like a team, and it can create a divide between who is special and working on the "new" thing versus who is stuck maintaining the "old" thing. Context switching between things has an overhead, including from a management perspective.

↑People && ↓Scope: Consolidation

See Chapter 7.

↑People && =Scope: Growth

Adding staff to teams to support growth or address pain points created by under-staffing is a normal part of a growing organization.

WHAT TO CONSIDER

As you add people to a team, things need to adjust. You can handle more work (good), but more work being done can put strain on processes that previously worked (like code review or releases).

WHAT TO WORRY ABOUT

When teams have been understaffed—especially if it has gone on for a while—it can feel dramatic and result in overhiring. It's important to be realistic about what is really needed and to ensure there's enough work—including suitable levels of onboarding work—for the people you add.

↑People && ↑Scope: High Growth!

In a period of hypergrowth, teams may both add people and grow in scope to meet the evolving needs of the organization.

WHAT TO CONSIDER

This stage is exciting, but it's very easy to incur management debt. Additionally, as the mission of the team expands, it's easy to lose clarity and prioritize poorly. There's more about this in Section 4, "Self-Improving Teams".

WHAT TO WORRY ABOUT

It's easy for this phase of growth to be wasteful and inefficient. If your hiring process is poor, you can easily mishire and then have to spend a lot of time correcting your mistakes. Often, people-management roles fall behind IC hiring, resulting in managers having way too many direct reports while they frantically try to hire other managers and then navigate the learning curve of managing managers... As people frantically take on more responsibilities to meet the growth need, they may get promoted above their capabilities, making it hard for them to be successful in the long term. We talk more about all of these things in Section 3, "Scaling Teams".

Reading List

Allen, David. *Getting Things Done: The Art of Stress-Free Productivity*. Penguin, 2015.

Arbinger Institute. *Leadership and Self-Deception: Getting Out of the Box*. Berrett-Koehler, 2002.

Babcock, Linda, and Sara Laschever. *Women Don't Ask: Negotiation and the Gender Divide*. Princeton University Press, 2021.

Bar-David, Sharone. *Trust Your Canary: Every Leader's Guide to Taming Workplace Incivility*. Bar-David Consulting, 2015.

Bridges, William, and Susan Bridges. *Managing Transitions: Making the Most of Change*. Hachette Books, 2017.

Brown, Brené. *Dare to Lead*. Random House, 2018.

Buckingham, Marcus, and Ashley Goodall. *Nine Lies About Work: A Freethinking Leader's Guide to the Real World*. Harvard Business Review Press, 2019.

Buckingham, Marcus, and Curt Coffman. *First, Break All the Rules: What the World's Greatest Managers Do Differently*. Gallup Press, 2014.

Buckingham, Marcus, and Donald O. Clifton. *Now, Discover Your Strengths*. Gallup Press, 2001.

Cabane, Olivia Fox. *The Charisma Myth: How Anyone Can Master the Art and Science of Personal Magnetism*. Penguin, 2013.

Chamine, Shirzad. *Positive Intelligence: Why Only 20% of Teams and Individuals Achieve Their True Potential and How You Can Achieve Yours*. Greenleaf Book Group Press, 2012.

DeMarco, Tom. *Slack: Getting Past Burnout, Busywork, and the Myth of Total Efficiency*. Crown, 2002.

Fine, Cordelia. *Delusions of Gender: How Our Minds, Society, and Neurosexism Create Difference*. W. W. Norton, 2011.

Fournier, Camille. *The Manager's Path: A Guide for Tech Leaders Navigating Growth and Change.* O'Reilly, 2017.

Gladwell, Malcolm. *Blink: The Power of Thinking Without Thinking.* Little, Brown, 2007.

Goldsmith, Marshall. *What Got You Here Won't Get You There: How Successful People Become Even More Successful.* Profile, 2010.

Gottlieb, Lori. *Maybe You Should Talk to Someone: A Therapist, Her Therapist, and Our Lives Revealed.* Houghton Mifflin Harcourt, 2019.

Grant, Adam. *Give and Take: Why Helping Others Drives Our Success.* Penguin, 2014.

———. *Hidden Potential: The Science of Achieving Greater Things.* Penguin, 2023.

Hastings, Reed, and Erin Meyer. *No Rules Rules: Netflix and the Culture of Reinvention.* Penguin, 2020.

Heath, Chip, and Dan Heath. *Switch: How to Change When Change Is Hard.* Crown, 2010.

Huston, Therese. *Let's Talk: Making Effective Feedback Your Superpower.* Penguin, 2021.

King, Michelle P. *The Fix: Overcome the Invisible Barriers that Are Holding Women Back at Work.* Atria Books, 2020.

Maslach, Christina, and Michael P. Leiter. *The Truth About Burnout: How Organizations Cause Personal Stress and What to Do About It.* Wiley, 2008.

McDowell, Gayle Laakmann. *Cracking the Coding Interview: 150 Programming Interview Questions and Solutions.* CareerCup, 2011.

Meadows, Donella H. *Thinking in Systems: A Primer.* Chelsea Green Publishing, 2008.

Miller, Jo. *Women of Influence: 9 Steps to Build Your Brand, Establish Your Legacy, and Thrive.* McGraw Hill, 2019.

Morgan, Adam, and Mark Barden. *A Beautiful Constraint: How to Transform Your Limitations into Advantages, and Why It's Everyone's Business.* Wiley, 2015.

Pink, Daniel H. *Drive: The Surprising Truth About What Motivates Us.* Penguin, 2011.

Reilly, Tanya. *The Staff Engineer's Path.* O'Reilly, 2022.

Reynolds, Marcia. *Coach the Person, Not the Problem: A Guide to Using Reflective Inquiry.* Berrett-Koehler, 2020.

Rubin, Gretchen. *Better Than Before: What I Learned About Making and Breaking Habits.* Crown, 2015.

————. *The Four Tendencies: The Indispensable Personality Profiles That Reveal How to Make Your Life Better (and Other People's Lives Better, Too)*. Harmony/Rodale, 2017.

Rumelt, Richard. *Good Strategy/Bad Strategy: The Difference and Why It Matters*. Crown, 2011.

Schein, Edgar H. *Helping: How to Offer, Give, and Receive Help*. Berrett-Koehler, 2011.

Steele, Claude. *Whistling Vivaldi: How Stereotypes Affect Us and What We Can Do*. W. W. Norton, 2011.

Stone, Douglas, and Sheila Heen. *Thanks for the Feedback: The Science and Art of Receiving Feedback Well*. Penguin, 2015.

Vanderkam, Laura. *168 Hours: You Have More Time Than You Think*. Penguin, 2011.

————. *Off the Clock: Feel Less Busy While Getting More Done*. Little, Brown, 2018.

Watkins, Michael D. *The First 90 Days: Proven Strategies for Getting Up to Speed Faster and Smarter*. Harvard Business Review Press, 2013.

Index

About the Author

Cate Huston has had a global career in technology, focused on scaling teams and leaders within fast-growing engineering organizations. She currently works as an Engineering Director at DuckDuckGo and previously worked at Automattic (the company behind WordPress.com), both globally distributed companies. She is also a Co-Active trained engineering leadership coach, a Limited Partner in the Acquired Wisdom VC fund, and an advisor at Glowforge.

Cate holds a BSc (hons) in computer science from the University of Edinburgh and started her career as a software engineer at Google. She speaks internationally and writes regularly for Quartz, and her writing has also appeared in *Lifehacker*, *The Daily Beast*, *LeadDev*, and in the collections *97 Things Every Engineering Manager Should Know*, *Living by the Code*, and *The Architecture of Open Source Applications*. Cate appeared in the books *The Manager's Path*, *Coders*, and *Brotopia* and as a tech industry expert on *The Today Show* (US) and the Channel 4 News (UK).

Colophon

The original art is by Susan Thompson. The map is an SVG-designed file cut on a Cricut using wood veneer and hand-painted and assembled. The compass is an SVG-designed file cut on a Cricut using 80-lb Canson-colored paper and assembled. The cover fonts are Guardian Sans Condensed-Medium, Semibold, and Regular, Guardian Sans Regular, and Gilroy Semibold. The text fonts are Minion Pro and Scala Pro; the heading and sidebar font is Benton Sans.

Printed in the USA
CPSIA information can be obtained
at www.ICGtesting.com
LVHW081618230824
789095LV00001B/34